Water Treatment with
New Nanomaterials

Water Treatment with New Nanomaterials

Editor

Shahin Homaeigohar

MDPI • Basel • Beijing • Wuhan • Barcelona • Belgrade • Manchester • Tokyo • Cluj • Tianjin

Editor
Shahin Homaeigohar
Aalto University
Finland

Editorial Office
MDPI
St. Alban-Anlage 66
4052 Basel, Switzerland

This is a reprint of articles from the Special Issue published online in the open access journal *Water* (ISSN 2073-4441) (available at: https://www.mdpi.com/journal/water/special_issues/Water_Treatment_New_Nanomaterials).

For citation purposes, cite each article independently as indicated on the article page online and as indicated below:

LastName, A.A.; LastName, B.B.; LastName, C.C. Article Title. *Journal Name* **Year**, *Article Number*, Page Range.

ISBN 978-3-03936-810-5 (Hbk)
ISBN 978-3-03936-811-2 (PDF)

Cover image courtesy of Shahin Homaeigohar.

Contents

About the Editor

Shahin Homaeigohar is Assistant Professor and Lecturer in Biomedical Engineering at the University of Dundee, UK. Prior to his current position, he was a Research Scientist and Postdoctoral Fellow at University of Erlangen-Nuremberg, Germany; Aalto University, Finland; and Helmholtz Centre for Materials and Coastal Research (HZG), Germany. He has authored over 43 publications (26 as first author), one book chapter, and 2 books (one reprinted by ISPO) in addition to being granted 4 patents (1 US patent) and receiving numerous recognitions (Falling Walls Lab Award, Kajal Mallick Memorial Award, Publon's Top Peer Reviewer Award, Marie Curie Seal of Excellence, and Distinguished Researcher Student Award). Some of his works have appeared as frontispieces and cover images of prominent journals, including *Advanced Materials, Advanced Functional Materials, Advanced Optical Materials, Advanced Materials Interfaces, Nano Today, and Materials Today*. He is a recipient of the prestigious European Marie Skłodowska Curie Actions (MSCA) grant, the German Helmholtz-DAAD PhD fellowship, numerous travel grants, and an Academy of Finland mobility grant. Dr. Homaeigohar has also actively served as a reviewer for high-ranked journals, with his efforts culminating in him being awarded Top Peer Reviewer by Publons. He is a member of the Topic Editorial Board of *Nanomaterials* and has served as Guest Editor for *Water, International Journal of Molecular Sciences, and Materials*.

Editorial

Water Treatment with New Nanomaterials

Shahin Homaeigohar [†]

Nanochemistry and Nanoengineering, Department of Chemistry and Materials Science,
School of Chemical Engineering, Aalto University, Kemistintie 1, 00076 Aalto, Finland;
Shahin.homaeigohar@fau.de
† Current address: Institute of Biomaterials, Department of Materials Science and Engineering,
University of Erlangen-Nuremberg, 91058 Erlangen, Germany.

Received: 10 May 2020; Accepted: 25 May 2020; Published: 25 May 2020

Abstract: The studies introduced in this special issue aim to provide a state-of-the-art vision for nanomaterials-based technology that could profit the water treatment industry. Given the expanding crisis of water shortages across the world, this perspective is invaluable and of paramount importance. No doubt, as the environmental challenges are going to be more complicated and to extend to as-yet unconsidered areas, we need to upgrade our facilities and knowledge to address them properly. Nanomaterials are indeed promising building blocks for such advanced technologies that enable them to purify water streams from complex pollutants in an energy, cost and time-effective manner. The focus of the (review and original research) articles collected in this issue is on various kinds of nanomaterials made of carbon, polymer, metal, and metal oxides (magnetic and photocatalyst), that are employed for adsorption and photodegradation of heavy metals and organic pollutants, respectively. Here, I briefly review the insights given in these precious studies and suggest new directions for future research in this field.

Keywords: nanomaterial; water treatment; adsorption; membrane; photocatalysis; nanohybrids; ecotoxicology

---

1. Introduction

While water shortage across the world are threatening the well-being of the human community, emerging advanced technologies targeted to address this challenge are promising. In this regard, nanomaterials have played a crucial role and offered new opportunities for the construction of permeable and selective membranes and adsorbents. Such features are of paramount importance, particularly given the limited available energy resources. In this issue, it was aimed to cover new researches dealing with water treatment based on nanomaterials made of polymer, metal, composite, ceramic (metal oxide), carbon, etc., that are shaped in various dimensionalities such as particle (0D), fiber (1D), and film (2D–3D).

In 2018, the special issue of "Water treatment with new nanomaterials" was introduced to the research community actively working on this subject all around the world. Following the call, within the course of running this issue, several articles were submitted to the editorial office, thereof seven were accepted for publication. These included five original research papers and two review papers. In the next sections, the published studies are briefly summarized.

2. Review Papers

Both reviews published in this issue give an overview of the potentials of nanomaterials for different water treatment approaches in a critical manner. Homaeigohar et al. [1] classified the nanomaterials studied for water purification based on the technologies benefitting from them, as nanosized adsorbents, nanomembranes, nanophotocatalysts, etc. They reviewed the relevant state-of-the-art approaches while pointing out to the potential bottlenecks. As an important aspect of this study, they discussed the fate of nanomaterials released into soil and water and their potential adverse impacts on the respective biota. As they report, according to the "ISI Web of Knowledge" statistics on the subjects discussed in this review, i.e., "Nanomaterials for Water Treatment" and "Environmental Impacts of Nanomaterials," 222 and 63 respective articles have been published from 2010 to 2020. These numbers imply that the former topic has been paid more attention during the past ten years, while the second topic has been taken less seriously in the research community. Given the potential impacts and risks of nanomaterials on the environment, there is a need to allocate more time and budget to research on this critical topic.

Yaqoob et al. [2] reviewed various classes of nanomaterials studied for the sake of water treatment depending on their function. A particular emphasis has been given on nanophotocatalysts and their potential applications and operation mechanism, nano/micromotors, nanomembranes, and nanosorbents. The merits of this review could be its futuristic outlook on these classes of nanotechnolgies and its critical view while highlighting the available potentials.

3. Original Research Papers

Homaeigohar and Elbahri [3] reported on the synthesis of carbon buckypaper adsorbents made from amphiphilic (oxygenated amorphous carbon (a-CO_x)/graphite (G)) nanofilaments. Such a buckypaper is able to dynamically adsorb and separate organic biomolecules (i.e., urease enzyme) from water. Bearing in mind the dynamic nature of the test, this adsorbent has shown to be optimally efficient in adsorption of the enzyme (88%) and at the same time to be notably permeable to water (4750 $L \cdot h^{-1} m^{-2} bar^{-1}$); thus, assuring selectivity and permeability required for a low energy consuming water treatment.

Heydari et al. [4] studied and compared the relative efficiency of four groups of supported titanium dioxide (TiO_2) that were utilized for the purpose of photodegradation of 2,4-dichlorophenoxyacetic acid (2,4-D) in Killex®, a commercially available herbicide. These four groups included floating TiO_2 spheres, an anodized TiO_2 plate with nanotube arrays, anodized TiO_2 mesh with a 3D nanotube structure, and electro-photocatalysis using the anodized TiO_2 mesh. Their energy consumption upon exposure to an ultraviolet light-emitting diode (UV-LED) light source at λ = 365 nm in a semi-passive mode was employed as an indicator for photocatalysis efficiency. Based on the authors' results, 3D nanotubes of TiO_2 surrounding the mesh enable a higher photodegradation efficiency compared to 1D arrays on the anodized plate.

Zhang and Wang [5] synthesized nanoscale zero-valent iron (nZVI) particles via pulse electrodeposition and loaded them on the surface of biomass activated carbon (BC). The BC-nZVI composite was employed as an adsorbent for the removal of methyl orange (MO) from water. The adsorption test implied that the adsorbent is able to remove 97.94% MO from water in one hour. This removal percentage can be optimally enhanced with a higher temperature, a larger BC-nZVI dosage, and a lower initial concentration of MO under neutral pH.

Campagnolo et al. [6] developed an Au/ZnO hybrid photocatalyst mounted on a poly (methyl methacrylate) (PMMA) fibrous substrate in a sequential manner. The fabrication process comprised the thermal conversion of the ZnO precursor already incorporated into PMMA electrospun fibers and then dipping of ZnO/PMMA fibers into Au precursor solution and heating. The as-formed hierarchical heterogeneous photocatalyst was utilized for photodegradation of organic water pollutants, specifically methylene blue (MB) and bisphenol A (BPA), under UV light.

Ying et al. [7] synthesized a nanocomposite adsorbent system comprising meso- and microporous tire-derived-carbon support loaded with iron oxide nanoparticles. This nanocomposite adsorbent selectively adsorbed Se(IV) ions from simulated wastewater. The adsorption kinetics of this nanocomposite adsorbent was assessed in a fixed-bed setting and after several column runs by inductively coupled plasma-optical emission spectroscopy (ICP-OES), quantifying the concentration changes over time. Fitting the obtained data to a pseudo-second-order rate law, equilibrium values were determined that alongside the effluent concentration data could help the calculation of reaction constants and column coefficients through the Adams-Bohart model. As the authors believe, their findings allow the utilization of this nanocomposite adsorbent for fixed-bed column systems and enable further research based on mixed pollutants in real wastewater.

4. Outlook and Future Directions

Nanomaterials are promising building blocks for the next generation of water purification technologies and water pollution control systems. To expedite the realization of the nano-based water treatment approaches, there are still several milestones that need to be reached:

(1) Next generation of nanoadsorbents: Nanomaterials can be employed in the development of nanoadsorbents with different surface functionalities that optimally capture polar and non-polar pollutants from water. By using nanophotocatalysts, adsorption is coupled with photodecomposition. Therefore, the adsorbed organic pollutant is decomposed to harmless byproducts and the previously occupied surface becomes free for a subsequent adsorption/photodecomposition cycle. In relevance to the photocatalysis based water treatment, it is crucial to hamper hole-electron recombination in the photocatalyst and also to replace UV light with solar visible light as the main stimulus for the process. This latter target assures energy efficiency and more extensive usability of photocatalysis for water treatment. A critical bottleneck for the nanoparticulate adsorbents is their aggregation tendency and difficult recovery. To address these problems, they could be deposited on nanofiber substrates. Accordingly, not only is the aggregation level notably alleviated and the recovery of the nanoparticles is facilitated, but also their high availability to the external water medium is preserved.

(2) Next generation of nanomembranes: Nanomaterials can be also used as the building blocks of nanostructured membranes. Particularly, electrospun nanofibers and graphene nanosheets have been widely investigated in the last decade for such a purpose. Electrospun nanofiber mats offer remarkable potential for size exclusion and also adsorption of water pollutants. Owing to their adjustable pore size, extraordinary porosity, and open porous structure, they enable energy-efficient, thus economical, water treatment process. For this reason, they have been appealing for the construction of state-of-the-art ultra- (UF) and nanofiltration (NF) membranes as a permeable, solid support for the selective layer. Despite such merits, nanofibrous membranes have not yet been industrially realized. This issue might stem from the lack of proper and reliable testing of such membranes. Nanofibrous membranes must be tested over a long time period, and under the chemical, thermal, and mechanical conditions commonly found in practice and with real wastewater models. Unrealistically, at the lab scale, the nanofibrous membranes are typically tested in the presence of only one type of pollutant and co-existence of other dye, ionic, or organic pollutants, as found in real wastewater, is ignored. Graphene membranes i.e., those comprising mono-/few-layer graphene nanosheets also belong to the next generation of water membranes. They theoretically can offer remarkable water permeability, while providing the ionic selectivity that is comparable to classic NF and ideally reverse osmosis (RO) membranes. However, their potentials for water treatment have been predominantly proven theoretically rather than experimentally, and practical employment of such membranes would take time.

(3) Energy efficiency and scalability: There are currently several barriers against the widespread commercial use of nanomaterials for water treatment. These include technical shortcomings relating to scale-up and integration of nanomaterials into water purification technology, safety, cost, and energy effectiveness. For example, TiO_2 nanoparticles and carbon nanotubes (CNTs) are of the most largely investigated nanomaterials for dye adsorption. Yet, their toxicity and costly production method involving high temperature and pressure are indeed discouraging for industrialization. Aside from that, TiO_2 nanoparticles should be UV irradiated to photodecompose the dye molecules that raises the costs of the process. Therefore, as a prospect, the application of nanomaterials for water treatment is justified in case they can be produced at large amounts with reasonable costs, proportional to different classes of wastewaters.

(4)   Sustainable and ecofriendly nanomaterials: The nanomaterials being used in the development of micro-, ultra-, and nanofiltration membranes can potentially be freed into water upon exposure of the membrane to aggressive water streams with intricate stress patterns. Thus, as an important priority, nanoparticles must be firmly stabilized on/in the membrane structure through physicochemical treatments. Moreover, nontoxic materials should be considered that are environmentally less challenging. In this regard, the new generation of nature-derived nanomaterials such as cellulose nanomaterials could be promising. Additionally, it is of utmost importance to develop synthesis and processing methods that are green and minimally involve hazardous chemicals. Short term investigations have implied that some nanomaterials are safe to human beings, plants, and animals. However, there is no guarantee for their long-term safety. Therefore, development of state-of-the-art water treatment systems based on nanomaterials should be pursued with caution. From the technological point of view, it is also crucial to properly design these systems so that they minimally release nanomaterials into the environment.

Funding: This research received no external funding.

Acknowledgments: The reviewers who kindly reviewed the submissions to this special issue and provided precious scientific inputs to the submitted articles are gratefully acknowledged.

Conflicts of Interest: The authors declare no conflict of interest.

References

1.   Ghadimi, M.; Zangenehtabar, S.; Homaeigohar, S. An Overview of the Water Remediation Potential of Nanomaterials and Their Ecotoxicological Impacts. *Water* **2020**, *12*, 1150. [CrossRef]

2.   Yaqoob, A.A.; Parveen, T.; Umar, K.; Mohamad Ibrahim, M.N. Role of nanomaterials in the treatment of wastewater: A review. *Water* **2020**, *12*, 495. [CrossRef]

3.   Homaeigohar, S.; Elbahri, M. An Amphiphilic, Graphitic Buckypaper Capturing Enzyme Biomolecules from Water. *Water* **2019**, *11*, 2. [CrossRef]

4.   Heydari, G.; Hollman, J.; Achari, G.; Langford, C.H. Comparative study of four TiO_2-based photocatalysts to degrade 2, 4-D in a semi-passive system. *Water* **2019**, *11*, 621. [CrossRef]

5.   Zhang, B.; Wang, D. Preparation of Biomass Activated Carbon Supported Nanoscale Zero-Valent Iron (Nzvi) and Its Application in Decolorization of Methyl Orange from Aqueous Solution. *Water* **2019**, *11*, 1671. [CrossRef]

6.   Campagnolo, L.; Lauciello, S.; Athanassiou, A.; Fragouli, D. Au/ZnO hybrid nanostructures on electrospun polymeric mats for improved photocatalytic degradation of organic pollutants. *Water* **2019**, *11*, 1787. [CrossRef]

7.   Ying, A.; Evans, S.F.; Tsouris, C.; Paranthaman, M.P. Magnetic Sorbent for the Removal of Selenium (IV) from Simulated Industrial Wastewaters: Determination of Column Kinetic Parameters. *Water* **2020**, *12*, 1234. [CrossRef]

Article

Magnetic Sorbent for the Removal of Selenium(IV) from Simulated Industrial Wastewaters: Determination of Column Kinetic Parameters

Andrew Ying [1,†]**, Samuel F. Evans** [1,2,†]**, Costas Tsouris** [3] **and M. Parans Paranthaman** [1,2,*]

1   Chemical Sciences Division, Oak Ridge National Laboratory, Oak Ridge, TN 37831, USA;
    andrewsying@berkeley.edu (A.Y.); sevans48@vols.utk.edu (S.F.E.)
2   The Bredesen Center for Interdisciplinary Research and Graduate Education, The University of Tennessee,
    Knoxville, TN 37996, USA
3   Energy and Transportation Science Division, Oak Ridge National Laboratory, Oak Ridge, TN 37831, USA;
    tsourisc@ornl.gov
*   Correspondence: paranthamanm@ornl.gov
†   Contributed equally.

Received: 1 April 2020; Accepted: 22 April 2020; Published: 26 April 2020

Abstract: A novel meso- and microporous tire-derived-carbon support with magnetic iron oxide nanoparticle adsorbents that selectively adsorbs Se(IV) ions from simulated contaminated water has been developed. In this work, the physicochemical characteristics of the composite adsorbent are characterized with respect to porosity and surface area, chemical composition, and microstructure morphology. The kinetics of this composite adsorbent in a fixed-bed setting has been determined. Several column runs were conducted and analyzed by inductively coupled plasma-optical emission spectroscopy (ICP-OES) to determine the concentration gradient vs time. These results were then fit to a pseudo-second order rate law to obtain equilibrium values. Combining calculated equilibrium values with effluent concentration data, enabled the application of the Adams–Bohart model to determine reaction constants and column coefficients. Column parameters obtained from different flow rates and fittings of the Adams–Bohart model were remarkably consistent. These findings enable the application of this sorbent to fixed-bed column systems and opens up further research into mixed pollutants tests with real wastewater and scaling of selenium pollutant removal.

Keywords: selenium removal; wastewater purification; nanoadsorbents; carbon magnetic iron oxide particles; bench scale column extraction; column kinetics

---

1. Introduction

Mining, fossil, and petrochemical operations often produce significant quantities of wastewater with high concentrations of toxic metals such as selenium, arsenic, and lead [1–4]. The millions of gallons of wastewater produced from industrial activities present an economic, legal, and technological challenge as the water must be treated before being stored or reintroduced back into the environment [5–8]. Previous research has explored the use of composite materials derived from sources of industrial waste, such as tires and iron chloride [9], as adsorbents that can affordably and sustainably reduce selenium concentrations to EPA standards of ≤50 ppb. These materials can be easily integrated into an industrial purification process by flowing wastewater through a fixed bed of composite adsorbents in a continuous-flow system.

Ion uptake rates of adsorbents vary when scaling from batch adsorption to continuous column tests due to transport effects. Fixed beds can experience uneven flow distribution, z-axis concentration gradients, mass transport limitations due to the presence of inert media, and other mass transport limitations. However, fixed-bed columns are of simple design and perform effectively in the removal of low concentration pollutants [10]. Kinetics models, such as the Adams–Bohart model, have been developed to model the sorbent behavior in continuous settings [11]. Employing such models to understand the behavior at a small scale can ease transition to full-scale industrial processes. By quantifying the maximum adsorption capacity, rate of adsorption, and breakthrough curve (a plot of concentration of the adsorbate in the effluent vs. time) [12], the operation of adsorption columns can be optimized. Rate law and kinetic constants can also be used to normalize columns to account for varying volumes, masses, and flow rates.

A variety of materials are known to adsorb selenium species effectively including mesoporous materials [13], graphene based composites [14], and a variety of metal oxides [15], layered-double hydroxides [16], metal-organic frameworks (MOFs) [17], MgO nanosheets [18], carbon nanotubes [19], and bioremediation-based methods [20–22]. Other methods, such as mineralization [23], capacitive deionization, and catalyzed reduction [24] are also known but not the focus of the current work. As separation materials are developed, they need to be aligned with a more sustainable future in mind. There are twelve principles of circular chemistry of particular importance [25]. In this work, the adsorbent material addressed two concerns in this area, namely the collection and use of waste and the optimization of resource efficiency. The sorbent was synthesized from waste tire material and iron chloride, $FeCl_3$, both of which are waste products from different sectors. Additionally, the sorbent was able to be regenerated and reused, decreasing the need for large amounts of adsorbent material.

A tire-derived carbon has recently been demonstrated in our laboratory as an effective support for iron oxide magnetic adsorbents for selectively removing Se(IV) ions from contaminated water [9]. While it is most effective in a fluidized bed, due to the ease of removal of the magnetized particles, determining its viability in fixed-bed systems is still of importance. This is due to the lower cost, simplicity, and predictable behavior in the removal of low concentration pollutants. Herein, the physicochemical characteristics of this composite adsorbent were characterized, and its kinetic behavior with respect to Se(IV) uptake in a fixed-bed setting was determined and analyzed to inform its effective use.

2. Experimental Methods

2.1. Carbon Supported Magnetic Nanoparticle Adsorbents (C-MNA) Synthesis

Tire derived carbon was synthesized as described in previous work [9]. To enhance its activity, it was ground with potassium hydroxide pellets (Sigma Aldrich, >85%) in a 1:4 weight ratio. The mixture was then placed in a furnace and heated at a 10 °C/min ramp rate to 800 °C, where it was kept under nitrogen atmosphere for 1.5 h. The mixture was then removed, cooled to ambient temperature, and neutralized with 3M hydrochloric acid (EMD Millipore).

C-MNA was synthesized by suspending 0.5 g of activated carbon in 120 mL of deionized water containing 13 mmol of iron sulfate, $FeSO_4 \cdot 7H_2O$ (Sigma Aldrich, ≥99.0%) and 15.6 mmol $FeCl_3 \cdot 6H_2O$ (Sigma Aldrich, ≥97.0%). The mixture was then sonicated for 5 min and stirred at 70 °C for 1 h. 3M sodium hydroxide solution was added in excess to maximize iron oxide nanoparticle (FeNP) precipitation. C-MNA was then vacuum filtered and separated from the excess iron chloride mother liquor via magnetic filtration and washing with DI water.

2.2. Column Set Up

A solution containing 5 ppm Se(IV) at pH 5 was prepared by dissolving sodium selenite (Sigma Aldrich, 99%) into deionized water. The pH was adjusted with dilute hydrochloric acid. As demonstrated in Figure 1, the column was prepared by using a standard jacketed borosilicate glass column with a radius of 0.5 cm and a bed height of 13 cm, lining the bottom with approximately 2.5-cm height of cotton, adding sand containing 5 weight % of C-MNA, and finally adding 2.5-cm height of pure sand on the top. A peristaltic pump was used to deliver the Se(IV) solution at varying flow rates. The effluent was collected manually in increments of 9.5 mL and sampled for analysis via inductively coupled plasma optical emission spectroscopy (ICP-OES) [26]. The column was prepared for trials by flowing 20 mL of dilute hydrochloric acid of pH 5 through the column before adding the Se(IV) solution. After the adsorbent was fully saturated with Se(IV), the column was flushed with 1 M sodium hydroxide solution of pH 11 to desorb Se(IV). This demonstrates that the adsorbents can be recycled as shown in Figure 1.

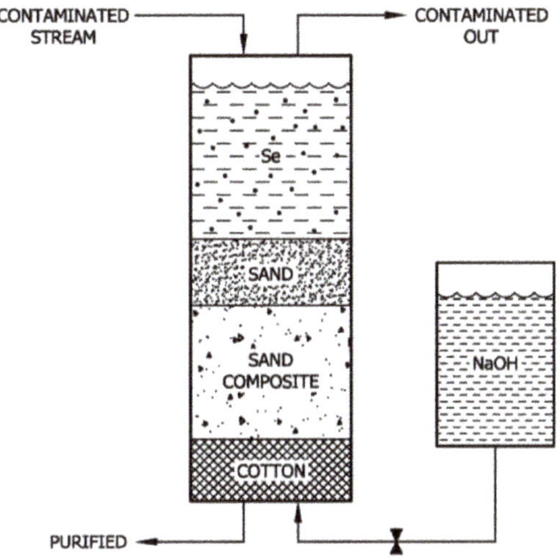

Figure 1. Schematic of the bench scale column set up. After Se (IV) removal from the contaminated stream, 1M NaOH solution can be flushed into the column to desorb Se (IV) in the form of sodium selenite by exiting through the top of the column.

2.3. Characterization

X-ray powder diffraction (XRD) patterns were collected using a PANalytical Empyrean instrument (Malvern, UK) with a Cu K_α radiation. All data were processed with HighScore Plus (Malvern, United Kingdom). Brunauer-Emmett-Teller (BET) surface areas and pore-size distributions were determined from nitrogen adsorption isotherms at 77 K using Autosorb-1 from Quantachrome (Anton Paar GmbH, Austria). The pore-size distributions and pore volumes were calculated from the DFT/Monte Carlo method using the QSDFT adsorption branch model. ICP-OES compositional analysis was performed to determine selenite removal from the solution using a Thermo Fisher iCAP Model 7400 ICP-OES Duo. The ICP-OES has a minimum selenium detection limit of 0.77 µg/L at wavelength 196 nm. During measurements, the linear standard curve had an R^2 value of 1 and a limit of detection of 0.049 ppm. A Zeiss Merlin VP scanning electron microscopy (SEM) (White Plains, NY, USA) operated at 3 kV and a Hitachi HD-2300A scanning transmission electron microscope (STEM) with a field emission source operated at 200 kV in bright-field imaging mode at a 2.1 Å resolution, were used to characterize the surface morphologies of the samples.

2.4. Kinetic Model

The kinetic equations were derived from the kinetic rate law equation for pseudo-second order reactions, as a variety of carbon based materials display this general adsorption behavior [27,28]. Equation (1) assumes that the adsorption capacity is correlated to the number of active sites on the surface.

$$\frac{dq_t}{dt} = k(q_e - q_t)^2 \tag{1}$$

In this equation, q_e and q_t refer to the equilibrium concentration and the concentration at time t, respectively. The k parameter is the kinetic constant. With the boundary conditions of $t_0 = 0$, $t = t$, $q_0 = q(t = 0) = 0$, and $q(t) = q_t$, a solution was derived and rearranged to a linear form as seen in Equations (2)–(5). Equation (5) is then used to fit experimental data.

$$\frac{1}{q_e - q_t} - \frac{1}{q_e} = kt \tag{2}$$

$$\frac{q_t}{q_e^2 - q_e q_t} = kt \tag{3}$$

$$q_t = \frac{ktq_e^2}{1 + ktq_e} \tag{4}$$

$$\frac{1}{q_t} = \left(\frac{1}{kq_e^2}\right)\frac{1}{t} + \frac{1}{q_e} \tag{5}$$

The derivation of the Adams–Bohart model starts with a modified chemical rate law expression as shown in Equation (6) [29].

$$\frac{\partial q}{\partial t} = k_{AB}C(q_0 - q) \tag{6}$$

The analytical solution provided by Adams and Bohart is shown in Equations (7) and (8) [30].

$$\frac{C}{C_0} = \frac{exp(\alpha)}{exp(\alpha) + exp(\beta) - 1} \tag{7}$$

$$\alpha = k_{AB}C_0\left(t - \frac{Z}{v}\right); \beta = \frac{k_{AB}\rho_p q_0 Z}{v}\frac{1 - \varepsilon}{\varepsilon} \tag{8}$$

Here, the following simplifications are made [28]:

1. $exp(\alpha), exp(\beta) \gg 1$

    The quantity 1 at the denominator of Equation (7) is thus regarded as insignificant.
2. $t \gg \frac{Z}{v}$

    Since the time of the experiment far outweighs the residence time, the residence-time term can be ignored.

The following definitions are also used:

1. $\rho_P q_0 (1 - \varepsilon) = N_0$
2. $\varepsilon v = u$

Equation (7) is then simplified into the recognizable form in Equation (9) [12]. Although this equation is used to calculate the breakthrough time, or the time before the effluent concentration exceeds acceptable levels [12], the Adams–Bohart equation can be used to predict changes in the effluent concentration over time.

$$ln\left(\frac{C_o}{C_B} - 1\right) = \frac{k_{AB}N_o Z}{u} - k_{AB}C_o t_B \tag{9}$$

Equation (9) can be rewritten with the following adjustments as seen in Equation (10).

$$\frac{N_0 Z}{u} = \frac{q_o M}{Q}$$

We can rewrite u as $\frac{Q}{A}$ and then rewrite $N_0 Z A$ as $q_o M$ since both reduce to the amount of solute adsorbed in the system.

$$ln\left(\frac{C_o}{C_B} - 1\right) = \frac{k_{AB} q_o M}{Q} - k_{AB} C_o t_B \tag{10}$$

3. Results and Discussion

The C-MNA adsorbent utilized in this study was fully characterized, including SEM, EDS, and magnetic property measurements in previous work [9]. XRD patterns of the column mixture of adsorbent and inert packing material shown in Figure 2 indicates the presence of both C-MNA and sand in the mixed material. SEM images of sand, C-MNA and the composite mixture are shown in Figure 3. SEM images of C-MNA and the porous carbon surfaces are clearly seen. Additionally, the large sand particles are mixed with smaller C-MNA particles homogenously in the mixture. Finally, a BET analysis was performed to determine the surface area of the different materials as another assurance of C-MNA mixing with the inert packing material. In Figure 4, surface areas of 6.8 m^2/g for sand, 638 m^2/g for C-MNA, and 45.5 m^2/g for the mixture were determined, confirming that the high surface area C-MNA was mixed with inert sand packing material.

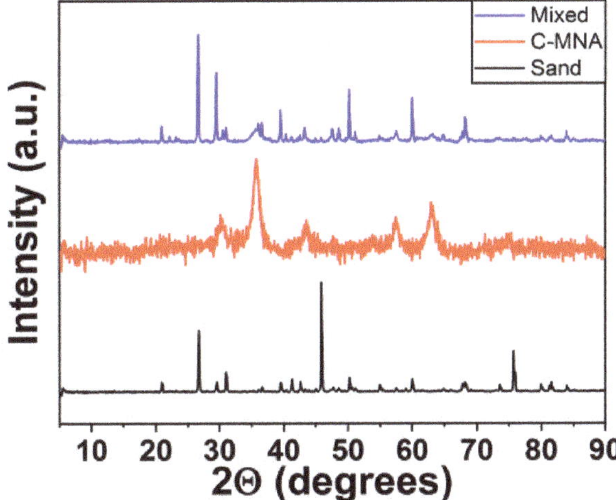

Figure 2. X-ray powder diffraction (XRD) patterns of column materials sand, C-MNA, and a mixture of the two.

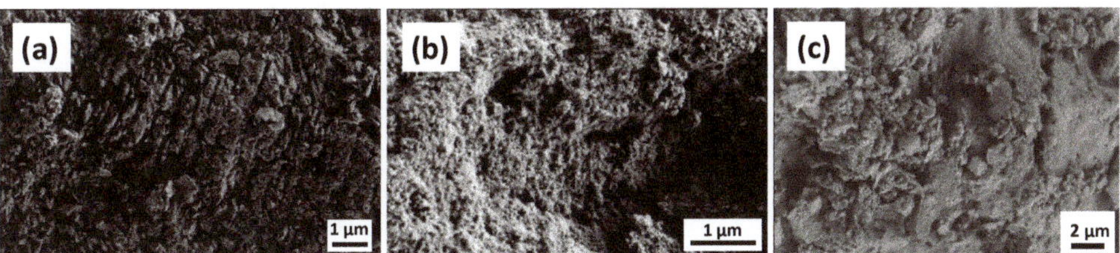

Figure 3. SEM images of (**a**) sand; (**b**) C-MNA adsorbent material; and (**c**) mixture of sand and C-MNA composite material.

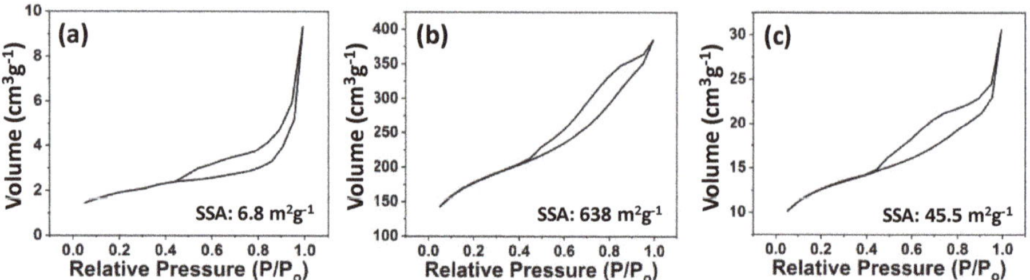

Figure 4. BET analysis of (**a**) sand; (**b**) C-MNA adsorbent material; and (**c**) mixture of sand and C-MNA used in the column.

In continuous flow systems, kinetics data are normalized for comparison to the residence time. The residence time, or the total time that the solution is in contact with the sorbent, is calculated by dividing the total bed volume by the volumetric flow rate.

$$Residence\ Time = \frac{Bed\ Volume}{Volumetric\ Flow\ Rate} \tag{11}$$

By expressing kinetics relative to residence time, we can make sure that the effects of bed volume size or flow rate are not omitted. In our column runs, we assumed that there was good radial mixing and no axial dispersion. As seen in Figure 5, the composite adsorbent steadily adsorbed Se(IV) until reaching a plateau between 0.4 and 0.5 mg Se per g adsorbent, suggesting monolayer formation on the carbon surface of the adsorbent [29]. Subsequent adsorption is thought to be a result of mass transfer of Se(IV) into the mesopores of the carbon support structures and formation of a multilayer on the surface of the iron nanoparticles, since the adsorption profile fits the two stage adsorption characteristic of a Type IV adsorption isotherm [19]. At a higher flow rate of Se(IV) solution, the initial rate of Se(IV) adsorption is higher than Se(IV) adsorption at a lower flow rate because there is a greater concentration gradient between the bulk concentration and the concentration adsorbed [27], which means a greater driving force for mass transfer.

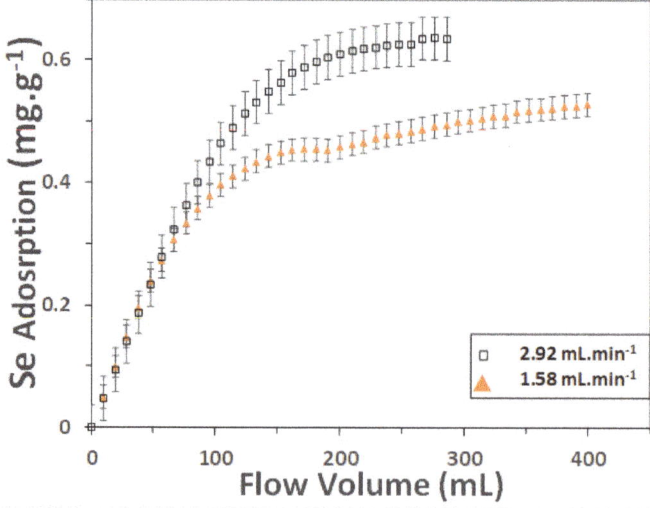

Figure 5. Se adsorption of column at differing flow rates to determine breakthrough times.

Sand was mixed with adsorbent because the small grain size of the adsorbent caused packing issues. Previous studies used sand as an inert medium to house sorbents [26]. However, sand is known to physisorb selenium as well as other contaminants [31,32]. To quantify the adsorption capacity of the sand, a column was filled with pure sand (still including the cotton layer at the bottom), and water containing 5 ppm Se(IV) at a pH of 5 was flowed through the column. The concentration of the effluent was compared to the initial concentration to calculate how much Se(IV) was adsorbed per gram of sand. At a flow rate of 2.92 mL/min and a pH value of 5, the adsorption capacity was low (6.8 µg/g), as shown in Figure 6, compared to the total of ~500 µg/g in the presence of the composite adsorbent. Based on these tests, the adsorption of sand was determined to be negligible. Instead, the adsorption is attributed to the embedded FeNPs and surface adsorption on the carbon support. Although correction for Se(IV) adsorption by sand did not change fundamental conclusions about the reaction rate or adsorption behavior, the equilibrium concentrations and kinetics constants were affected. As a result, all data subsequently shown and discussed contain corrections made for sand adsorption.

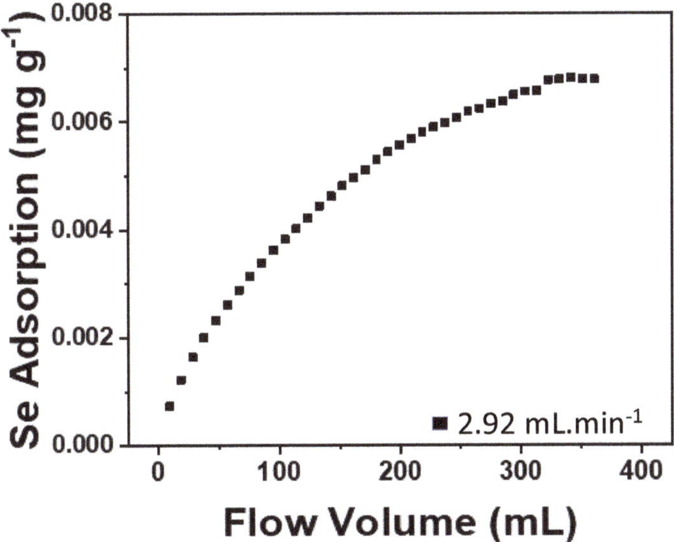

Figure 6. Adsorption of Se in sand as a control for inert column materials present.

3.1. Rate Law

To further elucidate the mechanism and kinetics of the composite adsorption of previously studied carbon and iron-based adsorbents, the adsorption rate was modeled using a second-order kinetic rate equation, as shown in Equation (7). From Figure 7, the adsorption kinetics is strongly consistent with pseudo-second order reactions and corroborates trends observed in kinetic data taken from other studies [33,34]. The equilibrium concentration and kinetic constant were also obtained from this model and subsequently used in Adams–Bohart kinetic model calculations (see Table 1).

Table 1. Equilibrium concentration and kinetic constants from 2nd order kinetic model fitting.

Residence Time (min)	Theoretical Se Uptake (mg g⁻¹)	Observed Se Uptake (mg g⁻¹)	Kinetic Constant (g mg⁻¹ min⁻¹)
3.25	3.305	0.513	0.001
6	0.724	0.407	0.013

Previous batch studies report the maximum adsorption concentration achieved was 1.14 mg/g [9]. When the sorbent is mixed with inert sand in a fixed-bed column, several factors could lead to a decrease in the observed uptake of Se. Channeling could restrict water flow to certain sections of the column, preventing the sorbent in that region from interacting with selenite. The inert medium could cover the mesopores in the composite sorbent, again restricting adsorbate flow to the iron nanoparticles housed within the mesopores. In addition, batch tests are performed with the adsorbent free flowing in solution, thus increasing the wettability of the sorbent in comparison to the fixed column. Finally, due to lower concentrations of the adsorbate near the bottom of the column, preventing the full utilization of the adsorbent for selenite removal, the adsorbent in that area could require extended time to reach equilibrium.

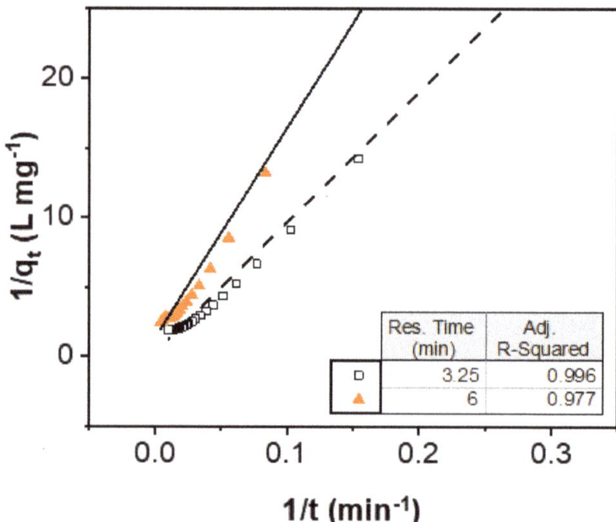

Figure 7. Linear regression fit of data to second order kinetics equation in agreement with expected pseudo-second order behavior.

3.2. Adams–Bohart Model

The Adams–Bohart model utilizes the following assumptions: (1) flow rate is constant, (2) absence of axial dispersion, (3) behavior matches the rectangle (irreversible) isotherm (i.e., highly favorable adsorption), and (4) adsorption rate follows second-order reaction kinetics [30]. The data were fit with a linear regression line as shown in Figure 8, and the rate constant (k_{BA}) was calculated from the m and b terms of the linear regression equation (y = mx + b).

The literature shows two forms (Equations (9) and (10)) of the Adams–Bohart model [11,30]. Although both models yield kinetic constants while maintaining dimensional homogeneity, they differ slightly in the method by which the kinetic constant is derived. One equation can be used to calculate the constants by equilibrium adsorption per unit volume, while the other can be used to determine the constants on a per unit mass basis. The two forms of the same model would be considered identical if the contents of the column were homogenized and uniform [30]. However, the presence of both an inert medium and the composite sorbent in our column means that local densities and adsorption capacities could vary. In our case, we chose to include the mass and volume of sand in our calculations in order to obtain more precise values. As discussed above, further tests indicated that the sand did play a minor role in the adsorption of Se(IV).

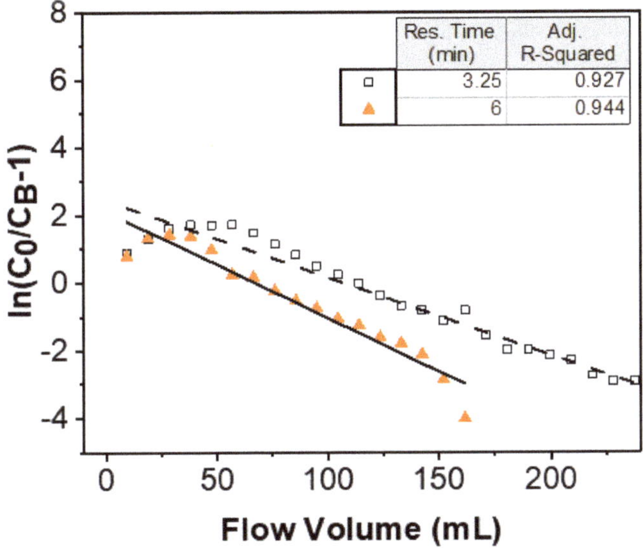

Figure 8. Linear regression fit of data to Adams–Bohart model. Values from these regressions were used in the calculation of kinetic constants in Tables 2 and 3.

Table 2. Kinetic constants calculated from the slope.

Residence Time (min)	Equation Used	k_{AB} (mL/mg/min)
3.25	9	0.011
	10	0.013
6	9	0.009
	10	0.009

Table 3. Kinetic constants calculated from the y-intercept.

Residence Time (min)	Equation Used	k_{AB} (mL/mg/min)
3.25	9	0.002
	10	2.167
6	9	0.005
	10	4.661

Linear regression fits were performed on the data in Figure 8, providing two equations to determine kinetic parameters from. For a resonance time of 3.5 min and 6 min, the regressions were y = −0.0502x + 2.1309 and y = −0.0617x + 2.4508, respectively. Since the rate constant appears in both the m and b terms of the linear regression equation, each fitting could yield two rate constants. Between these two terms the slope (m) produced more consistent parameter values, making them more useful for the predictions of the fixed-bed behavior [30]. A better agreement with the observed column adsorption behavior was seen by only including the initial adsorption of the column. This is reflective of the constraints of the model, as the Adams–Bohart model is typically used to depict breakthrough curves rather than overall adsorption curves [12]. As seen in Tables 2 and 3, many of the kinetic constants derived from the m and b components across all forms of the Adams–Bohart model and all residence times were similar. Only the values calculated on a per unit mass basis (Equation (10)) deviated significantly, potentially due to variance in equilibrium concentration calculations or sand adsorption behavior. The differing rate constants from the m and b terms potentially originate from multiple kinds of adsorption taking place or multiple reactions occurring. Another possibility is the existence of a pseudo-solution which can be eliminated by comparing the kinetic rate constants from both equations and over both runs. When compared to values obtained from experimental data, the theoretical values calculated from concentration measurements resulted in rate constants that displayed a small initial deviation from the experimentally obtained values that subsequently decreased as the column reached full saturation.

4. Conclusions

Column kinetics were determined by first analyzing the rate law to obtain equilibrium concentrations. Subsequently, this data was fitted to the Adams–Bohart model, enabling the calculation of critical data on column kinetics of C-MNA adsorbents in a fixed-bed system. Since the data displayed strong correlation to a pseudo-second order linear regression fit, the equilibrium concentrations could be applied to an Adams–Bohart providing a better understanding of column behavior. Within a continuous-flow setting, the Adams–Bohart model predicts column behavior with reasonable accuracy, given certain limitations. Given the mixture of semi-inert sand and composite adsorbent, calculations were found to be more accurate when the mass and volume of the sand were included in the model. Given the lack of literature on the Se(IV) adsorption behavior in a fixed-bed column, this analysis provides valuable insight as corporations and governments continue to require new technologies for waste and contaminated water processing. This work provides insights to enable further work on mixed pollutant removal in addition to industrial scaling of this column.

Finally, although magnetism has not been taken advantage of in the current work, it can be used to create macroporosity in a sorbent bed and thereby increase mass transfer, while reducing pressure drop. It is well known that magnetically stabilized beds can have the benefits of fixed beds, in terms of simplicity and high separation efficiency, and fluidized beds with respect to high mass-transfer rates and low pressure drop. Future work will be focused on removing Se(IV) from real wastewater using magnetically stabilized sorbent beds with magnetic sorbent based on tire-derived carbon.

Author Contributions: Conceptualization, A.Y., S.F.E., C.T. and M.P.P.; methodology, A.Y., S.F.E. and C.T.; formal analysis, A.Y., S.F.E. and C.T.; resources, M.P.P.; data curation, A.Y. and S.F.E.; writing—original draft preparation, A.Y., S.F.E.; writing—review and editing, C.T., M.P.P.; supervision, M.P.P.; funding acquisition, M.P.P. All authors have read and agreed to the published version of the manuscript.

Funding: The synthesis of carbon composite materials was sponsored by the U.S. Department of Energy, Office of Science, Office of Basic Energy Sciences, Materials Sciences and Engineering Division. This work was supported in part by the U.S. Department of Energy, Office of Science, Office of Workforce Development for Teachers and Scientists (WDTS) under the Science Undergraduate Laboratory Internship (SULI) program. S.F.E. is grateful for a fellowship from the Bredesen Center. We would like to thank Dr. Rich Lee, RJ Lee Group for providing tire derived carbon.

Acknowledgments: This manuscript has been authored by UT-Battelle LLC under Contract No. DE-AC05-00OR 22725 with the U.S. Department of Energy. The United States Government retains and the publisher, by accepting the article for publication, acknowledges that the United States Government retains a non-exclusive, paid-up, irrevocable, world-wide license to publish or reproduce the published form of this manuscript, or allow others to do so, for United States Government purposes. The Department of Energy will provide public access to these results of federally sponsored research in accordance with the DOE Public Access Plan (http://energy.gov/downloads/doe-public-access-plan).

Conflicts of Interest: The authors declare no conflict of interest. The funders had no role in the design of the study; in the collection, analyses, or interpretation of data; in the writing of the manuscript, or in the decision to publish the results.

Abbreviations

Relevant Variables

k	kinetic constant (g mg^{-1} min^{-1})
q_e	equilibrium concentration (mg g^{-1})
q_t	concentration at time t (mg g^{-1})
t	time (min)
C_B	breakthrough concentration (mg cm^{-3})
C	sorbate concentration in bulk (mg cm^{-3})
C_0	initial sorbate concentration in feed (mg cm^{-3})
k_{AB}	Adams–Bohart rate constant (cm^3 mg^{-1} s^{-1})
M	mass of adsorbent (g)
N_0	sorption capacity per unit volume of fixed bed (mg cm^{-3})
q	sorbate concentration in adsorbent (mg g^{-1})
q_0	sorption capacity per unit mass of adsorbent (mg g^{-1})
Q	flow rate (cm^3 s^{-1})
t_B	breakthrough time (s)
u	superficial velocity (cm s^{-1})
v	interstitial velocity (cm s^{-1})
V	volume of solution (mL)
Z	total bed depth (cm)
ε	column void fraction
ρ_P	apparent adsorbent density (g cm^{-3})

References

1. Liu, Y.-T.; Chen, T.-Y.; Mackebee, W.G.; Ruhl, L.; Vengosh, A.; Hsu-Kim, H. Selenium Speciation in Coal Ash Spilled at the Tennessee Valley Authority Kingston Site. *Environ. Sci. Technol.* **2013**, *47*, 14001–14009. [CrossRef]
2. Walls, S.J.; Jones, D.S.; Stojak, A.R.; Carriker, N.E. Ecological risk assessment for residual coal fly ash at Watts Bar Reservoir, Tennessee: Site setting and problem formulation. *Integr. Environ. Assess. Manag.* **2015**, *11*, 32–42. [CrossRef]
3. Khamkhash, A.; Srivastava, V.; Ghosh, T.; Akdogan, G.; Ganguli, R.; Aggarwal, S. Mining-related selenium contamination in Alaska, and the state of current knowledge. *Minerals* **2017**, *7*, 46. [CrossRef]
4. Lemly, A.D. Aquatic selenium pollution is a global environmental safety issue. *Ecotoxicol. Environ. Saf.* **2004**, *59*, 44–56. [CrossRef]
5. Grimalt, J.O.; Ferrer, M.; Macpherson, E. The mine tailing accident in Aznalcollar. *Sci. Total Environ.* **1999**, *242*, 3–11. [CrossRef]
6. Johnson, D.B. Chemical and microbiological characteristics of mineral spoils and drainage waters at abandoned coal and metal mines. *Water Air Soil Pollut. Focus* **2003**, *3*, 47–66. [CrossRef]
7. Deonarine, A.; Kolker, A.; Doughten, M.W. *Trace Elements in Coal Ash*; US Geological Survey: New York, NY, USA, 2015; pp. 2327–6932.
8. Hartuti, S.; Kambara, S.; Takeyama, A.; Kumabe, K.; Moritomi, H. Direct quantitative analysis of arsenic in coal fly ash. *J. Anal. Methods Chem.* **2012**, *2012*. [CrossRef]
9. Evans, S.F.; Ivancevic, M.R.; Yan, J.; Naskar, A.K.; Levine, A.M.; Lee, R.J.; Tsouris, C.; Paranthaman, M.P. Magnetic adsorbents for selective removal of selenite from contaminated water. *Sep. Sci. Technol.* **2019**, *54*, 2138–2146. [CrossRef]
10. Sze, M.F.F.; Lee, V.K.C.; McKay, G. Simplified fixed bed column model for adsorption of organic pollutants using tapered activated carbon columns. *Desalination* **2008**, *218*, 323–333. [CrossRef]
11. Aksu, Z.; Gönen, F. Biosorption of phenol by immobilized activated sludge in a continuous packed bed: Prediction of breakthrough curves. *Process Biochem.* **2004**, *39*, 599–613. [CrossRef]
12. Taniguchi, M.; Wang, K.; Gamo, T. *Land and Marine Hydrogeology*; Elsevier: Amsterdam, The Netherlands, 2003.
13. Awual, M.R.; Hasan, M.M.; Ihara, T.; Yaita, T. Mesoporous silica based novel conjugate adsorbent for efficient selenium (IV) detection and removal from water. *Microporous Mesoporous Mater.* **2014**, *197*, 331–338. [CrossRef]

14. Fu, Y.; Wang, J.; Liu, Q.; Zeng, H. Water-dispersible magnetic nanoparticle–graphene oxide composites for selenium removal. *Carbon* **2014**, *77*, 710–721. [CrossRef]
15. Sheha, R.; El-Shazly, E. Kinetics and equilibrium modeling of Se (IV) removal from aqueous solutions using metal oxides. *Chem. Eng. J.* **2010**, *160*, 63–71. [CrossRef]
16. Asiabi, H.; Yamini, Y.; Shamsayei, M. Highly selective and efficient removal of arsenic(V), chromium(VI) and selenium(VI) oxyanions by layered double hydroxide intercalated with zwitterionic glycine. *J. Hazard. Mater.* **2017**, *339*, 239–247. [CrossRef]
17. Howarth, A.J.; Katz, M.J.; Wang, T.C.; Platero-Prats, A.E.; Chapman, K.W.; Hupp, J.T.; Farha, O.K. High Efficiency Adsorption and Removal of Selenate and Selenite from Water Using Metal–Organic Frameworks. *J. Am. Chem. Soc.* **2015**, *137*, 7488–7494. [CrossRef]
18. Cui, W.; Li, P.; Wang, Z.; Zheng, S.; Zhang, Y. Adsorption study of selenium ions from aqueous solutions using MgO nanosheets synthesized by ultrasonic method. *J. Hazard. Mater.* **2018**, *341*, 268–276. [CrossRef]
19. Vilardi, G.; Mpouras, T.; Dermatas, D.; Verdone, N.; Polydera, A.; Di Palma, L. Nanomaterials application for heavy metals recovery from polluted water: The combination of nano zero-valent iron and carbon nanotubes. Competitive adsorption non-linear modeling. *Chemosphere* **2018**, *201*, 716–729. [CrossRef]
20. Khakpour, H.; Younesi, H.; Mohammadhosseini, M. Two-stage biosorption of selenium from aqueous solution using dried biomass of the baker's yeast Saccharomyces cerevisiae. *J. Environ. Chem. Eng.* **2014**, *2*, 532–542. [CrossRef]
21. Tuzen, M.; Sarı, A. Biosorption of selenium from aqueous solution by green algae (Cladophora hutchinsiae) biomass: Equilibrium, thermodynamic and kinetic studies. *Chem. Eng. J.* **2010**, *158*, 200–206. [CrossRef]
22. Nettem, K.; Almusallam, A.S. Equilibrium, Kinetic, and Thermodynamic Studies on the Biosorption of Selenium (IV) Ions onto Ganoderma Lucidum Biomass. *Sep. Sci. Technol.* **2013**, *48*, 2293–2301. [CrossRef]
23. Serrà, A.; Artal, R.; García-Amorós, J.; Sepúlveda, B.; Gómez, E.; Nogués, J.; Philippe, L. Hybrid Ni@ZnO@ZnS-Microalgae for Circular Economy: A Smart Route to the Efficient Integration of Solar Photocatalytic Water Decontamination and Bioethanol Production. *Adv. Sci.* **2020**, *7*, 1902447. [CrossRef]
24. Hill, C.M. *Review of Available Technologies for the Removal of Selenium from Water*; Final Report for North American Metals Council: Washington, DC, USA, June 2010.
25. Keijer, T.; Bakker, V.; Slootweg, J.C. Circular chemistry to enable a circular economy. *Nat. Chem.* **2019**, *11*, 190–195. [CrossRef]
26. Islam, M.T.; Saenz-Arana, R.; Hernandez, C.; Guinto, T.; Ahsan, M.A.; Bragg, D.T.; Wang, H.; Alvarado-Tenorio, B.; Noveron, J.C. Conversion of waste tire rubber into a high-capacity adsorbent for the removal of methylene blue, methyl orange, and tetracycline from water. *J. Environ. Chem. Eng.* **2018**, *6*, 3070–3082. [CrossRef]
27. Robati, D. Pseudo-second-order kinetic equations for modeling adsorption systems for removal of lead ions using multi-walled carbon nanotube. *J. Nanostruct. Chem.* **2013**, *3*, 55. [CrossRef]
28. Cooney David, O. *Adsorption Design for Wastewater Treatment*; CRC Press LLC: Boca Raton, FL, USA, 1999; pp. 9–20.
29. Ho, Y.-S.; McKay, G. Pseudo-second order model for sorption processes. *Process Biochem.* **1999**, *34*, 451–465. [CrossRef]
30. Chu, K.H. Fixed bed sorption: Setting the record straight on the Bohart–Adams and Thomas models. *J. Hazard. Mater.* **2010**, *177*, 1006–1012. [CrossRef] [PubMed]
31. Kent, D.; Davis, J.; Anderson, L.; Rea, B. Transport of chromium and selenium in a pristine sand and gravel aquifer: Role of adsorption processes. *Water Resour. Res.* **1995**, *31*, 1041–1050. [CrossRef]
32. Awan, M.A.; Qazi, I.A.; Khalid, I. Removal of heavy metals through adsorption using sand. *J. Environ. Sci.* **2003**, *15*, 413–416.
33. Maiti, A.; DasGupta, S.; Basu, J.K.; De, S. Batch and column study: Adsorption of arsenate using untreated laterite as adsorbent. *Ind. Eng. Chem. Res.* **2008**, *47*, 1620–1629. [CrossRef]
34. Tanhaei, B.; Ayati, A.; Lahtinen, M.; Sillanpää, M. Preparation and characterization of a novel chitosan/Al2O3/magnetite nanoparticles composite adsorbent for kinetic, thermodynamic and isotherm studies of Methyl Orange adsorption. *Chem. Eng. J.* **2015**, *259*, 1–10. [CrossRef]

Article

Au/ZnO Hybrid Nanostructures on Electrospun Polymeric Mats for Improved Photocatalytic Degradation of Organic Pollutants

Laura Campagnolo [1,2], Simone Lauciello [3], Athanassia Athanassiou [1] and Despina Fragouli [1,*]

[1] Smart Materials, Istituto Italiano di Tecnologia, Via Morego 30, 16163 Genova, Italy
[2] Dipartimento di Chimica e Chimica Industriale, Università degli Studi di Genova, Via Balbi 5, 16126 Genova, Italy
[3] Electron Microscopy Facility, Istituto Italiano di Tecnologia, Via Morego 30, 16163 Genova, Italy
* Correspondence: despina.fragouli@iit.it; Tel.: +39-010-289-6878

Received: 31 July 2019; Accepted: 22 August 2019; Published: 28 August 2019

Abstract: An innovative approach for the fabrication of hybrid photocatalysts on a solid porous polymeric system for the heterogeneous photocatalytic degradation of organic pollutants is herein presented. Specifically, gold/zinc oxide (Au/ZnO)-based porous nanocomposites are formed in situ by a two-step process. In the first step, branched ZnO nanostructures fixed on poly(methyl methacrylate) (PMMA) fibers are obtained upon the thermal conversion of zinc acetate-loaded PMMA electrospun mats. Subsequently, Au nanoparticles (NPs) are directly formed on the surface of the ZnO through an adsorption dipping process and thermal treatment. The effect of different concentrations of the Au ion solutions to the formation of Au/ZnO hybrids is investigated, proving that for 1 wt % of Au NPs with respect to the composite there is an effective metal–semiconductor interfacial interaction. As a result, a significant improvement of the photocatalytic performance of the ZnO/PMMA electrospun nanocomposite for the degradation of methylene blue (MB) and bisphenol A (BPA) under UV light is observed. Therefore, the proposed method can be used to prepare flexible fibrous composites characterized by a high surface area, flexibility, and light weight. These can be used for heterogeneous photocatalytic applications in water treatment, without the need of post treatment steps for their removal from the treated water which may restrict their wide applicability and cause secondary pollution.

Keywords: nanocomposite fibers; photocatalysis; mineralization; water remediation; organic pollutants

---

1. Introduction

The growing demand for clean water urges an improvement in commonly used methods for the wastewater management [1]. Depending on the type of water contamination, these conventional wastewater treatments can be physical, chemical, and biological processes [2]. However, these treatments present several limitations from a technological point of view, including the necessity for wastewater pre-treatment steps, secondary pollution, limited practicability, and often, low performance when emerging organic pollutants are involved [2–4]. For the removal of emerging and recalcitrant pollutants, new strategies have been proposed in combination with the available treatments. Among them, advanced oxidation processes (AOPs) can be used to ideally destroy organic pollutants from the contaminated water, thanks to the in situ production of strong oxidant species as the result of the interaction of the developed materials with solar, chemical, or other forms of energy [5–8].

Heterogeneous photocatalytic oxidation is one of the most promising AOPs for oxidizing organic pollutants into more biodegradable and less harmful compounds [5]. Typically, wide band gap, nano-, or micro-scale sized semiconductors are used as photocatalysts able to absorb incident phonons causing the subsequent formation of a separation charge with a conduction-band electron and a valence-band hole. The photo-excited carriers can generate hydroxyl radicals and other reactive oxygen species (ROS), which react with the organic molecules present in the contaminated water causing their degradation, assuring an eco-compatible process since the oxidation occurs in mild conditions [5].

The semiconductor particles are usually dispersed in the contaminated water to ensure a high exposed surface area in contact with the pollutants, and therefore, a high volumetric generation rate of ROS. However, the main limitation in their use for water remediation applications is the demand of energy and time consuming additional steps for their recovery from the treated water. To avoid these post-treatment steps, the photo-catalyst powders can be immobilized on suitable solid supports [9,10]. To this aim, polymeric supports have been recently introduced due to their low weight, flexibility, and easy conversion in different morphologies [11–16]. Among them, fibrous polymeric mats have high surface area, which allows an enhanced interaction with the polluted water, and their preparation process is relatively simple and easily scaled up [9,17].

For the fabrication of the functional semiconductor based polymeric materials, an innovative approach has been recently proposed, dealing with the direct synthesis of NPs inside different polymeric matrices through a solid-state thermally, chemically, or photo-induced reaction [13,17–19]. Compared to traditional mixing processes, this approach offers the advantageous formation of NPs homogeneously distributed not only in the whole volume of the polymer matrix, but also exposed on its surface. Furthermore, limitations such as the complex rheology of polymer/NP solutions, which can limit their electrospin-ability and therefore the formation of fibrous nanocomposites, can be avoided [13,17]. Among the diverse types of nanocomposites developed so far following this approach [13,17,20], the ZnO-based mats is a promising system for the water remediation through the photocatalytic approach. In fact, ZnO is of great interest not only because it is environmental friendly, non-toxic, of low-cost, and abundant, but also because of its optoelectronic properties [9,21–24]. In particular, the n-type ZnO is a direct wide band gap semiconductor (band gap energy, E_g = 3.37 eV), characterized by a high exciton binding energy (60 meV), which allows efficient excitonic emission even at room temperature [25]. However, due to the high photo-generated electron–hole pair recombination, the photocatalytic performance and, therefore, the utilization of the ZnO in wastewater purification, can be limited. For this reason, the enhancement of the separated electron–hole lifetime is one of the factors on which the efforts are focused in order to improve the ZnO photocatalytic performance [26–28]. This aspect can be achieved through various methods, such as the modification of the morphology of the ZnO nanostructures [22,29], the incorporation of metallic [30,31] and non-metallic dopants [32] in their crystalline structure, and the preparation of hybridized structures with other materials [21,22,33,34].

Concerning the last method, the combination of ZnO NPs with noble metal NPs has been recently considered an effective route to enhance the local density of states at the metal–semiconductor interface, prolonging thus the electron–hole separation lifetime [21,33]. Specifically, the use of Au NPs to modify the ZnO surface has been proven to improve the photocatalytic performance of the semiconductor due to the surface plasmonic resonance effect (which allows the establishment of the Schottky barrier simplifying the charge carrier separation), the absorption in the visible range of the electromagnetic spectrum, and the ROS generation [35,36].

Considering the above, the development of Au/ZnO-based PMMA fibrous mats is here presented and their application in the remediation of organic aqueous pollutants through the heterogeneous photocatalytic approach is thoroughly explored. Specifically, ZnO NPs are synthetized by a thermally-induced solid state reaction directly in the electrospun polymeric fibers, obtaining an easily handleable flexible solid porous material. The subsequent dipping of the PMMA/ZnO composite mats in Au precursor solutions of different concentrations and their thermal treatment result in the direct formation of Au NPs on the surface of the fibers and, therefore, on the exposed ZnO NPs, forming a high metal–semiconductor interfacial area. Different concentrations of the Au precursor solutions are used to explore the effect of the amount, size, and localization of the formed Au NPs on the

ZnO/PMMA fiber surface. Moreover, the photocatalytic performance of the different nanocomposite mats under UV light irradiation is investigated, focusing on the degradation of the model molecule MB and of the emerging pollutant BPA. In particular, it is demonstrated that the presence of Au NPs affects significantly the photocatalytic performance of the PMMA/ZnO mat for both the organic pollutants. Therefore, with the presented straightforward fabrication method, it is possible to form light-weight and flexible polymeric fiber mats of high surface area that contain functional NP combinations for advanced photocatalytic applications. This valuable alternative to the existing systems paves the way to an innovative route for the fabrication of multifunctional membranes for water purification without the need of costly and time consuming post treatment separation processes as in the case of powders.

2. Materials and Methods

2.1. Materials

PMMA (average Mw ~350 kDa), zinc acetate dihydrate ($Zn(CH_3CO_2)_2$, 99,999%), *N,N*-dimethylformamide (DMF, ≥99.8%), chloroauric acid trihydrate ($HAuCl_4 \cdot 3H_2O$, 99.999%), MB, BPA (≥99%), acetone (≥99.5%), ethanol (EtOH, ≥99.8%), sodium hydroxide (NaOH, ≥98%, anhydrous), hydrochloridric acid (HCl, 37%), and nitric acid (HNO_3, 70%), were purchased from Sigma Aldrich, Italy. All chemicals were used without any further purification.

2.2. Preparation of PMMA/ZnO Composite Mats

The nanocomposite fibers were prepared as described elsewhere [17]. In brief, 1.5 g of PMMA were added in 10 mL of DMF and left under stirring at 50 °C until the complete dissolution; then 1.1 g (c.a. 40 wt %) of $Zn(CH_3CO_2)_2$ were added to the polymer solution and stirred at 40 °C, until a clear solution is obtained. The electrospinning technique was then used in a vertical setup to obtain the $PMMA/Zn(CH_3CO_2)_2$ composite mats. The prepared solution was placed in a 10-mL syringe with a 22 G needle and the flow rate was set to 600 $\mu L \cdot h^{-1}$. The voltage applied to the syringe needle was 23 kV and the grounded collector was kept at a distance of 16 cm from the needle. The electrospun fibers were then dried to remove the eventual residual solvent for 8 h under dynamic vacuum. The in situ growth of the ZnO NPs on the electrospun mats was achieved by placing the electrospun composite mats in a convection oven for 48 h at 110 °C, through the thermal decomposition of $Zn(CH_3CO_2)_2$.

The obtained PMMA/ZnO composite mats were then washed in 10 mL of a H_2O/EtOH mixture (8:2 *v/v*) for 24 h and dried at 60 °C for 4 h. Following the methodology described to Section 2.4, it is defined that the total zinc loss during this washing step was around 1.170 ± 0.160 wt % of the total zinc. This can be attributed either to the loss of ZnO NPs not perfectly fixed on the surface of the fibers, or, also, to the water-soluble zinc acetate, which was not fully converted during the synthesis.

2.3. Preparation of PMMA/ZnO-Au Composite Mats

Three PMMA/ZnO composite mats (40 mg each) were dipped in 20 mL of a H_2O/EtOH mixture (8:2 *v/v*) of $HAuCl_4 \cdot 3H_2O$, which was previously neutralized to pH 6.5 by $NaOH_{aq}$ 0.1 M for 24 h in dark under gentle stirring. The aqueous-based $HAuCl_4$ $3H_2O$ solution was prepared at different initial concentrations of gold precursor—1.5 mM, 3 mM, and 4.6 mM—to explore the effect of the different amounts of Au on the photocatalytic performance of the PMMA/ZnO-Au mats. The composite mats were then collected and washed three times with water. The reduction of the gold ions was obtained by the subsequent thermal treatment of the mats at 60 °C for 4 h, obtaining three PMMA/ZnO-Au samples named PMMA/ZnO-Au1, PMMA/ZnO-Au3, and PMMA/ZnO-Au6, respectively. Each sample was then divided in four similar pieces of 10 mg each for further analysis.

2.4. Characterization

The fiber morphology was investigated by a scanning electron microscope (SEM, JEOL JSM 6490LA, Tokyo, Japan) and a high-resolution scanning electron microscope (HR-SEM, JEOL JSM

7500FA, Tokyo, Japan), equipped with a cold field emission gun applying an accelerating voltage of 15 kV. The specimens were previously coated with a 10-nm-thick carbon layer by a carbon coater (Emitech K950X, Quorum Technologies Ltd., East Susse, UK). The morphology of the ZnO and Au NPs in the composite mats was investigated by a transmission electron microscope (TEM, JEOL JEM 1400 Plus, Tokyo, Japan) operating at an acceleration voltage of 120 kV. To do so, the polymeric fibers were dissolved in acetone and the suspension was sonicated and centrifuged in order to separate the NPs from the polymer. Then, 30 µL of the suspension were deposited on a copper grid (CF300-Cu-UL, Electron Microscopy Science, USA) for further investigation. High-angle annular dark field scanning TEM (HAADF-STEM) images and energy dispersive X-ray spectroscopy (EDS) maps were acquired with a high-resolution FEI Tecnai G2 F20 (Oregon, USA) equipped with a Schottky field emission gun (FEG). The average size and the size distribution of the fibers and NPs were explored using the open source Fiji/ImageJ software.

The X-ray diffraction (XRD) analysis was performed on a PANalytical Empyrean X-ray diffractometer with a 1.8 kW CuKα ceramic X-ray tube (λ = 1.5418 Å) and a PIXcel^{3D} area detector (2×2 mm^2), operating at 45 kV and 40 mA. The diffractograms were recorded in parallel-beam geometry and symmetric reflection mode, for a 2θ range from 25° to 80°, with a step time of 383.67 s, and a step size of 0.026°. The average crystallite size was calculated by using the Debye–Scherrer equation (details in Supplementary Materials). The zinc and the gold content in the composite mats was obtained by ICP-OES analysis (iCAP 6300, Thermo, USA) of the digested solid samples. Specifically, 2.5 mL of aqua regia (HCl/HNO$_3$, 3:1) were added to 2.5 mg of the composite fibers and the solid degradation reaction was performed by using a microwave digestion system (MARS Xpress, CEM) at 180 °C for 15 min. The samples were then diluted with milliQ water up to 50 mL and filtered through polytetrafluoroethylene (PTFE) syringe filters (diameter 15 mm, pore size 0.45 µm, Sartorius). In addition, to evaluate the loss of ZnO and Au NPs from the composite mats, the ICP-OES analysis was performed on the liquids after the photocatalytic degradation process. The samples (125 µL) were previously treated with 2.5 mL of aqua regia, and then they were diluted up to 25 mL with milliQ water and filtered through PTFE syringe filters (diameter 15 mm, pore size 0.45 µm, Sartorius, Germany).

The optical properties of the nanocomposite fibers were evaluated by diffuse-reflectance measurements in the range between 200 and 800 nm, using a Varian Cary 6000i (Agilent, USA) UV-visible-NIR spectrophotometer equipped with integrating sphere. The energy bandgap (E$_g$) of ZnO and ZnO/Au NPs was then extrapolated by applying the Kubelka-Munk method to the reflectance results (see Supplementary Materials). The Raman spectra of the composite mats were acquired by a Raman LabRam HR800 (Horiba Jobin-Yvon Inc., France) spectrometer equipped with a built-in microscope with objectives $10 \times$ (NA 0.25) and $50 \times$ (NA 0.75), using the 632.8 nm He-Ne laser excitation. The experimental setup consists of a grating 600 lines·mm^{-1} with a spectral resolution of approximately 1 cm^{-1}.

2.5. Photocatalytic Performance

The photocatalytic performance of the developed materials was evaluated by immersing 10 mg of the mats in quartz cuvettes filled with 3 mL of MB or BPA aqueous solutions (0.0125 mM and 0.044 mM, respectively). To evaluate the adsorption of the organic molecules on the developed mats, experiments in dark conditions were performed overnight. For the photocatalytic experiments, the solutions containing the samples were placed at a distance of 10 cm under a UVA lamp emitting at a wavelength range from 315 nm to 400 nm (937 µW/cm^2 at 365 nm, 10 cm from source) connected to a climatic chamber (ICH 110 L, Memmert, Germany). The UVA irradiance (µW/cm^2) of the UV light source was measured by using a combined photo-radiometric probe (LP 471 P-A, Delta Ohm, Italy) at the fixed experimental distances between the source and the samples. Before starting the UV irradiation, the mats were maintained in the dark for 60 min. The decrease of the organic molecule concentrations was monitored by recording the UV-vis absorption spectra at specific time intervals.

The test was also performed for a reference sample, namely a quartz cuvette filled with the MB or BPA solution, in order to estimate the self-degradation of the pollutants under the UV light irradiation.

The photocatalytic degradation efficiency (*DE%*) was obtained from the Equation (1), while the first-order rate constant (k_1, min^{-1}) was calculated by fitting the experimental data of the first 5 h of irradiation with the pseudo-first order model expressed in Equation (2)

$$DE\% = \left(1 - \frac{C}{C_0}\right)\cdot 100 \tag{1}$$

$$\ln\left(\frac{C_0}{C}\right) = k_1 t \tag{2}$$

in which C_0 is the initial pollutant concentration and C is the concentration at t irradiation time. The organic molecule concentrations were determined by using two calibration curves with linear regression (R^2 values of 0.9983 and 0.9998 for MB and BPA, respectively). For the MB, the calibration curve was obtained by monitoring the absorbance of the MB at the wavelength of maximum absorbance (λ_{max} = 664 nm) with concentrations ranging from 0.00087 to 0.0125 mM. For the BPA calibration curve, the aqueous solution concentrations ranged from 0.0031 to 0.05 mM, and the absorbance variation was monitored at 276 nm.

Before and after the photocatalytic degradation of the organic molecules, the total organic carbon (TOC) of the solutions was analyzed by a membraPure uniTOC lab analyzer, Gemany. The total amount of organic bound carbon in the aqueous samples was measured by mean of a UV/reagent promoted oxidation into carbon dioxide (CO_2) which was then quantified by the non-dispersive infrared detector. After the photocatalytic degradation experiment, the MB and BPA aqueous solutions were collected from the quartz cuvettes and diluted with MilliQ water, in order to reach 15 mL in each case. To remove any possible solid residue, the aqueous solutions were centrifuged and then filtered using a polytetrafluoroethylene (PTFE) syringe filters (diameter 15 mm, pore size 0.20 µm, Sartorius, Germany). The reagent used was an acidic aqueous solution of sodium persulfate (10%$_{W/V}$) and phosphoric acid (3%$_{V/V}$). The inorganic carbon content was firstly removed and measured by the internal acidification step. The TOC value was used to evaluate the mineralization of the MB and of the BPA, and therefore, the efficiency of the photocatalytic degradation.

The stability of the composite mats was evaluated by repeating the photocatalytic experiments on the same samples over three cycles. The composite mats were washed before starting the subsequent cycle by dipping in clean water for 1 h under UV light irradiation (same conditions as previously described) in order to remove the possible adsorbed pollutants.

3. Results

The electrospun composite mats (photographs shown in Figure S1 in the Supplementary Materials) consisted of continuous defect-free polymeric fibers characterized by smooth surface, whose random distribution resulted in the formation of a porous material. Compared to the PMMA fibers (Figure S2), the PMMA/Zn(CH_3CO_2)$_2$ fibers exhibited a higher mean diameter with a broader size distribution, while their morphology remained rather similar, in accordance with our previous work [17]. In particular, even if the electrospinning operating parameters were preserved, the mean diameter of the polymeric fibers increased from 1.5 ± 0.3 to 6.2 ± 3.7 µm (for PMMA and PMMA/Zn(CH_3CO_2)$_2$, respectively). This difference was mainly due to the presence of the high content of Zn(CH_3CO_2)$_2$ which causes the enhancement of the viscosity of the solution (from 1.57 ± 0.05 Pa·s to 4.10 ± 0.15 Pa·s) [17].

After the thermal treatment of the PMMA/Zn(CH_3CO_2)$_2$ fibrous mats, ZnO NPs were formed, well distributed on the surface of the fibers but also in their bulk, as shown in Figure 1a and Figure S3. The content of ZnO NPs obtained after the zinc salt conversion was about 11.7 ± 0.9 wt.% with respect to the composite, as calculated by the ICP-OES analysis. Apart the ZnO NPs formation, it should be mentioned that the thermal treatment did not affect the overall size distribution of the polymeric fibers, which remained unchanged (Figure S3b). In order to obtain the ZnO/Au NPs hybrid structures on the polymeric fibers, the PMMA/ZnO composite mats were dipped in the gold precursor aqueous solutions of different initial concentrations. The neutralization of the acidic gold precursor solution, performed to avoid the solubilization of the amphoter ZnO NPs, determined the presence of $[AuCl_2(OH)_2]^-$ species which were favorably adsorbed on the positively charged surface of the ZnO NPs.

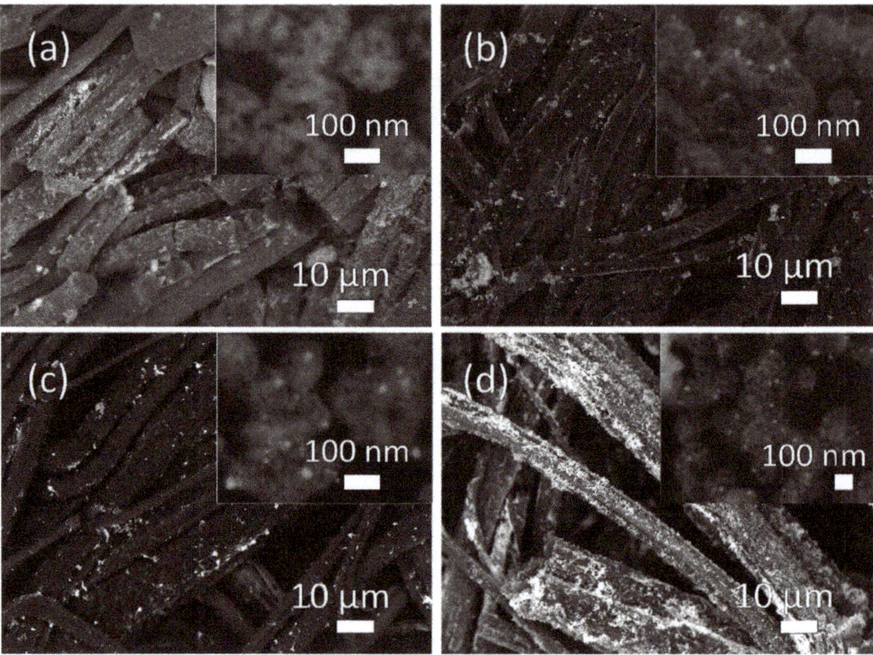

Figure 1. High-resolution scanning electron microscope (HRSEM) images of the (**a**) PMMA/ZnO, (**b**) PMMA/ZnO-Au1, (**c**) PMMA/ZnO-Au3, and (**d**) PMMA/ZnO-Au6 composite mats. Insets report the details on the surface of the different composite mats. PMMA: poly(methyl methacrylate).

The subsequent thermal treatment led to the formation of Au NPs on the surface of the fibers, which appeared not to be further affected by these steps. In fact, as shown in Figure 1b–d, the formed Au NPs appeared well distributed on the surface of the composite mats. The Au NPs were mainly attached on the surface of the ZnO NPs, as shown in the corresponding insets of the Figure 1b–d, while their size was not significantly affected by the initial concentration of the gold precursor solution. However, a higher initial concentration of the gold precursor solution led to an increase in the coverage of the composite fibers by the Au NPs, which, for the highest concentration (PMMA/ZnO-Au6), tended to form nanowires and aggregate on the surface of the composite mats. The amount of Au NPs formed on the composite mats was approximately 1%, 3%, and 6% wt. of Au with respect to the composite, for PMMA/ZnO-Au1, PMMA/ZnO-Au3, and PMMA/ZnO-Au6 respectively, as defined from the ICP-OES analysis.

As shown in Figure 2a,b, the formed ZnO NPs had a spherical branched structure with an average diameter of 135 ± 38 nm, which, in accordance with the SEM analysis, corresponded to the NPs grown on the surface of the fibers (Figure S3a). The selected area electron diffraction (SAED) pattern, inset of Figure 2a, confirmed the attribution of the NPs to ZnO of hexagonal wurtzite phase. In Figure S4a it is possible to observe also the presence of smaller irregular ZnO NPs. This second type of particles was formed in the internal part of the polymeric fibers (inset of Figure S3a), which restrained their growth [17].

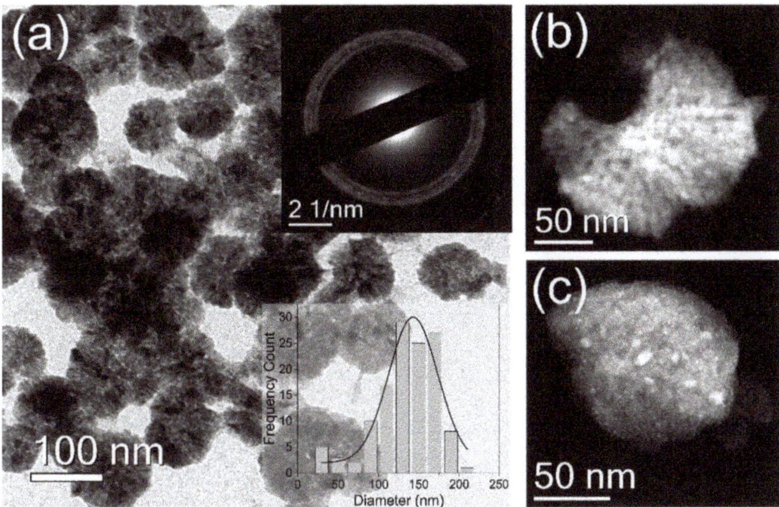

Figure 2. (**a**) Transmission electron microscope (TEM) image of the ZnO nanoparticles (NPs). The two insets show the selected area electron diffraction (SAED) pattern and the diameter size distribution of the NPs. Dark-field TEM images of (**b**) ZnO and of (**c**) ZnO decorated with Au NPs.

After the dipping in the gold precursor solutions of the PMMA/ZnO fibers and the subsequent heating process, small Au NPs were grown, mainly localized on the surface of the ZnO NPs, as shown in Figure 2c and in the EDS maps of Figure S5. In all the composite mats, the formed Au NPs had an average diameter of c.a. 8 ± 3 nm, 13 times smaller than the ZnO branched NPs. The lower mean diameter of the Au NPs compared to the ZnO allowed a high superficial contact between the two types of NPs and, therefore, an efficient charge separation.

The structures and the crystallites size of the ZnO NPs in the different composite mats were explored by XRD analysis (Figure 3). In all the composite mats, the ZnO NPs exhibited the characteristic diffraction peaks of the hexagonal wurtzite phase. In presence of Au NPs, a new diffraction peak appeared at 38°, which was assigned to the (111) crystal plane of the cubic Au phase. For the PMMA/ZnO-Au6 composite mat, it was also possible to observe an additional diffraction peak at 44°, related to the Au (002) lattice plane due to the higher amount of Au NPs on the surface compared to the other composite mats. The presence of Au did not cause shifts of the position of the ZnO diffraction peaks, and therefore the crystal structure of the ZnO NPs was not affected by the post-synthetic growth of the Au NPs. Furthermore, no extra phase or other peaks were observed, confirming the absence of crystalline impurities inside the samples. The average crystallite size of the ZnO NPs was approximated to 9 nm, as estimated by using the Debye–Scherrer equation.

The metal–semiconductor interface in the Au/ZnO NPs hybrid structure was investigated through diffuse reflectance and Raman spectroscopy. The diffuse reflectance spectra of the different composite mats are shown in Figure 4a. In all cases, it is possible to observe the characteristic absorption of the ZnO NPs in the UV region originated by the direct band gap transitions [26]. The presence of the Au NPs led to the addition of another absorption band with a maximum at 550 nm, attributed to the surface plasmon resonance of the Au NPs. The E_g of the ZnO was estimated by applying the Kubelka–Munk (KM) method to the data obtained by the reflectance spectra (details of the KM equation and graphical representation can be found in Supplementary Materials and Figure S6). The calculated E_g values demonstrate that the presence of the Au NPs did not cause any modification to the E_g of the ZnO present in the composite mats. In particular, the E_g was 3.25 ± 0.01 eV for the PMMA/ZnO, and 3.23 ± 0.02 eV, 3.23 ± 0.04 eV, and 3.25 ± 0.01 eV for the PMMA/ZnO-Au1, PMMA/ZnO-Au3, and PMMA/ZnO-Au6, respectively.

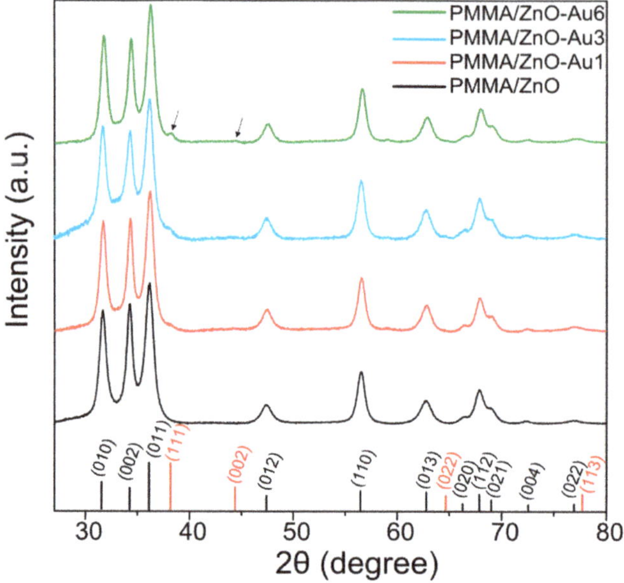

Figure 3. X-ray diffraction (XRD) patterns of the different composite mats. Stick reference pattern of ZnO (black) and of Au (red) are shown along the x-axis. The arrows indicate the detected Au diffraction peaks.

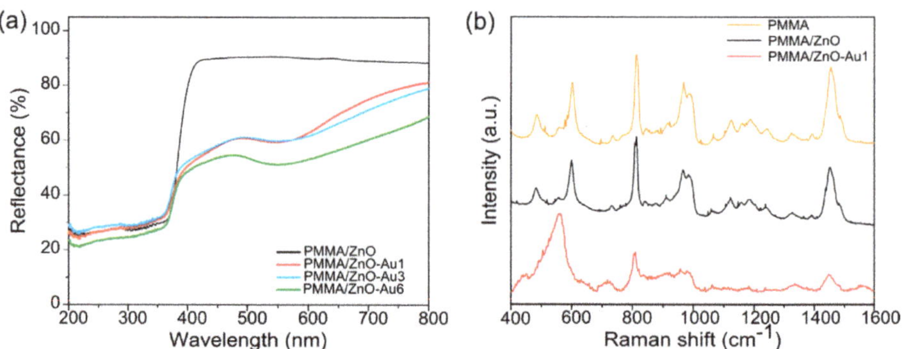

Figure 4. (**a**) Diffusive reflectance spectra of the developed composite mats. (**b**) Raman spectra of the polymeric fibers and of the PMMA/ZnO before and after the superficial modification with Au NPs.

The formation of the Au/ZnO NP hybrid structures was confirmed by Raman spectroscopy. Typically, the optical phonon modes predicted by the group theory for the hexagonal wurtzite ZnO structures are the polar Raman and infrared active modes A_1, E_1, the non-polar Raman active $2E_2$, and the silent $2B_1$ modes. The A_1 and E_1 modes are split in the transverse optical (TO) and longitudinal optical (LO) modes with different frequencies due to the manifestation of the macroscopic electric field of the LO phonons. The two E_2 are split in the high- and low-frequency modes (E_2^{high} and E_2^{low}) which are associated with the oxygen atoms and the zinc sublattice, respectively [37,38]. In the Raman spectrum of the PMMA/ZnO composite mat (Figure 4b), all the peaks observed were attributed to the PMMA [39], whose assignments are listed in Supplementary Materials (Table S1), while the Raman active modes of the wurtzite ZnO were not evident, possibly due to the presence of the polymer matrix and to the used experimental conditions [40]. It was possible to observe the characteristic modes of the ZnO only after the plasmonic enhancement provided by the contact with the Au NPs in the ZnO/Au hybrid systems. In fact, in the Raman spectrum of PMMA/ZnO-Au1 (Figure 4b), additional peaks appeared at 450 cm^{-1} and 560 cm^{-1}, which were respectively assigned to the E_2^{high} and A_1(LO) modes of the ZnO, proving the non-resonant surface enhanced Raman scattering of the ZnO optical modes in the proximity of Au NPs [37,41,42]. The Raman spectra of the PMMA/ZnO-Au3 and PMMA/ZnO-Au6 (Figure S7) show analogue enhancement of the Raman scattering of the ZnO optical modes.

The effect of different amounts of Au NPs on the photocatalytic degradation performance of the composite mats was then investigated using MB and BPA aqueous solutions. The MB is a cationic dye which can be found in wastewater deriving mainly from the textile, paper, and plastic industries. It is commonly adopted as a model pollutant for testing the photocatalytic performance of various materials under UV light irradiation. BPA is a recalcitrant aqueous pollutant classified as an endocrine disruptor, and is widely present in water due to its broad applications in the plastic industry [43]. In order to evaluate the adsorption of the organic pollutants on the fibrous composites, the mats were dipped in the organic pollutant solutions in dark conditions. As shown in Figure S8, there was a negligible MB adsorption on the fibers. The concentration of BPA slightly decreased over time in the presence of the PMMA/ZnO and of the PMMA/ZnO-Au with 1 wt % and 3 wt % Au contents. On the contrary, the PMMA/ZnO-Au6 showed a higher adsorption capacity compared to the other composites, adsorbing c.a. 39% of the initial BPA, due to the instauration of dispersive bonding interactions between the aromatic rings of the BPA and the gold surface, which was in a relatively high amount on the composite fibers [44,45].

Under UV irradiation (Figure 5), the concentration of both organic pollutants decreased in time only in the presence of the nanocomposite fibrous mats. In fact, their self-degradation was negligible, while for the pure PMMA fibrous mat there was no significant change in the concentration of the pollutants, confirming the absence of photocatalytic activity. When the composite mats were dipped in the solutions, it was possible to observe a more efficient photocatalytic degradation of the organic pollutants in the presence of the Au/ZnO hybrid structures.

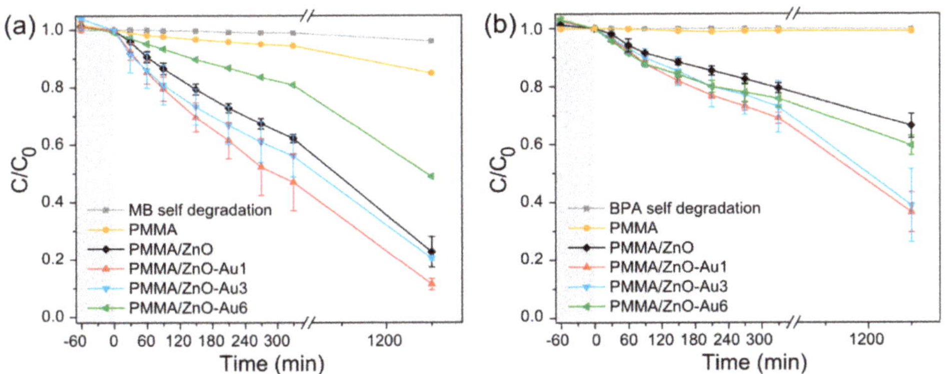

Figure 5. Photocatalytic degradation curves of (**a**) methylene blue (MB) and (**b**) bisphenol A (BPA) in presence of the composite mats under UV light. Before the UV irradiation, the solutions were kept in dark for 60 min in the presence of the composite mats. The self-degradation of MB and BPA under UV is also presented.

As shown in Figure 5a, the PMMA/ZnO mat was able to decolorize 77% of the initial MB solution while in the presence of the PMMA/ZnO-Au1, the efficiency reached 88% after 20 h of irradiation. By increasing the content of Au, the MB photocatalytic degradation tended to slightly decrease, reaching 80% for the PMMA/ZnO-Au3, while for the PMMA/ZnO-Au6, it was lower compared to that observed for the PMMA/ZnO (Figure 5a). The photocatalytic degradation curves of the pollutants followed a pseudo-first order kinetic model for all the developed composite mats (Figure S9), and the enhanced performance of the PMMA/ZnO-Au1 composite was also reflected on its k_1 for the first 5 h of irradiation. In fact, the photocatalytic degradation of MB with the PMMA/ZnO-Au1 reached the maximum value of k_1 2.33×10^{-3} min^{-1}, significantly higher than that obtained using the PMMA/ZnO, 1.44×10^{-3} min^{-1} and the other composite mats, as shown in Table S3 in the Supplementary Materials.

A similar behavior is also observed for the photocatalytic degradation of BPA. As shown in Figure 5b, the PMMA/ZnO-Au1 induced a 63.5% decrease of the initial BPA concentration after 20 h of UV irradiation, remarkably higher compared to the BPA reduction achieved with the PMMA/ZnO mat (*DE*% of 34%). For higher amounts of Au NPs, the photocatalytic degradation efficiency was around 61% and 41%, for the PMMA/ZnO-Au3 and PMMA/ZnO-Au6, respectively. The k_1 values calculated by applying the first order kinetics model on the experimental data were in accordance with the photodegradation efficiency results. The PMMA/ZnO-Au1 displayed a faster reaction rate (1.12×10^{-3} min^{-1}) compared to the PMMA/ZnO-Au3 and PMMA/ZnO-Au6 (0.96×10^{-3} and 0.82×10^{-3} min^{-1} respectively) and to the PMMA/ZnO (k_1 0.69×10^{-3} min^{-1}). Despite the results obtained for the MB, the PMMA/ZnO-Au6 composite mat displayed a more effective reduction of the initial BPA concentration compared to the PMMA/ZnO mat. However, this is not attributed to the photocatalytic degradation but to the high adsorption of the BPA on the surface of the Au NPs, as demonstrated with the adsorption experiment (Figure S8b).

For both the pollutants studied, the enhanced photocatalytic activity using the PMMA/ZnO-Au1 hybrid structures indicates that the specific ZnO-Au combination is favorable for the formation of the Schottky barrier at the metal–semiconductor interface, which enhances the charge carrier separation and therefore the ROS formation [21,46]. In fact, as observed in the SEM analysis, when the content of Au was low, the NPs were better distributed on the fibers without forming large aggregates, as in the case of higher amount of Au (PMMA/ZnO-Au6). The presence of Au aggregates on the ZnO NPs catalyst causes a decrease of their active surface exposed towards the organic pollutants and also a less effective light penetration due to a screening effect [22,28,47]. This result is also confirmed by previous studies; as a matter of fact, there is an optimal concentration of metallic NPs for the surface modification of the metal oxide semiconductors, beyond which the photocatalytic performance declines [46,48].

In order to evaluate the conversion of the organic pollutants into CO_2 and, therefore, the efficiency of the developed materials to mineralize the organic pollutants upon photocatalytic degradation, the mineralization of MB and BPA has been investigated by TOC studies of the solutions before and after the irradiation experiments. Based on the obtained results (Table S2 in Supplementary Materials), it can be assured that 60% of MB is mineralized in the presence of the PMMA/ZnO-Au1 mat upon UV irradiation, higher compared to the 45.5% obtained using the PMMA/ZnO. In the case of BPA, the PMMA/ZnO-Au1 was able to mineralize the 15% of the 63.5% photodegraded molecules, while no changes in the TOC were detected after using PMMA/ZnO. The lower mineralization of BPA compared to that obtained for MB can be ascribed to the recalcitrant nature of this organic pollutant. The effective removal of the BPA molecules is greatly challenging due to their complex aromatic structure, which, together with their low biodegradability, makes BPA one of the principal pollutants detected in the effluents of the wastewater treatment plants [49].

Since the PMMA/ZnO-Au1 mats show the best performance in terms of photocatalytic degradation of both MB and BPA organic pollutants, their stability is investigated by performing three consecutive irradiation cycles. The loss of Zn and Au from the composite mats was negligible after the first irradiation cycle, since about 0.022 ± 0.001 wt % of Zn with respect to the total Zn amount in the composites was detected in the liquids, while the content of Au was well below the instrument's detection limit. The overall photocatalytic performance of the PMMA/ZnO-Au1 was maintained unvaried for two consecutive cycles in terms of degradation efficiency of MB (Figure 6a), while, in the third cycle, the performance slightly decreased to 72.5%. This was also the case of the PMMA/ZnO sample (Figure S10a of Supplementary Materials) indicating the fact that this reduction in the performance can be possibly attributed to a low photo-corrosion effect of the ZnO [50]. In the case of BPA, the photocatalytic performance of the PMMA/ZnO-Au1 showed a more evident decrease after the second cycle compared to that observed for the MB, lowering further in the third cycle (Figure 6b). Despite this reduction, the degradation efficiency of the third cycle was higher than that reached for the PMMA/ZnO mat, in which only 10% of the BPA was photodegraded (Figure S10b). This aspect confirms the improvement of the catalytic activity of the mat due to the presence of 1% of Au NPs also in the removal of a recalcitrant aqueous pollutant.

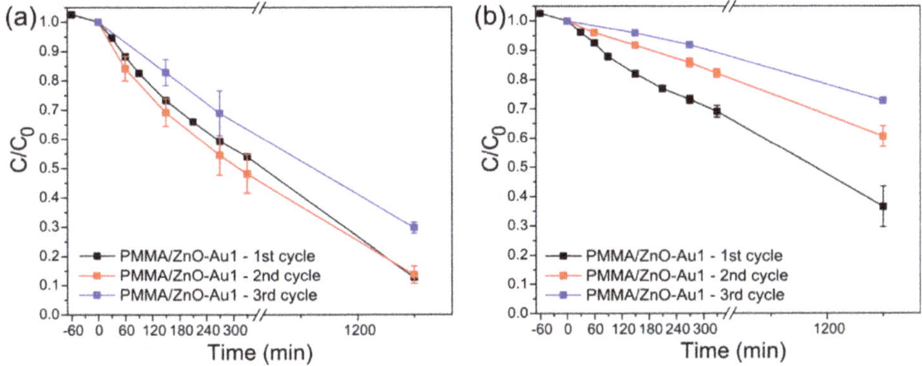

Figure 6. Photocatalytic degradation activity under UV light irradiation of (**a**) MB and (**b**) BPA for three consecutive cycles using the PMMA/ZnO-Au1.

4. Conclusions

The enhancement of the photocatalytic activity of the PMMA/ZnO composite mats was here investigated through a surface modification of the semiconductor nanostructures with Au NPs by means of a straightforward fabrication approach. The hybrid Au/ZnO nanostructures were fixed on a polymeric support through a thermally induced solid-state synthesis and a subsequent adsorption dipping process. Thanks to this accessible and scalable fabrication procedure, the Au/ZnO nanostructures were formed directly on the surface of electrospun polymeric fibers, obtaining a flexible, easily handleable, and light-weight porous system. The photoactive material was stable on the polymeric matrix, since its loss in water after 24 h was negligible. The presence of 1% of Au NPs on the PMMA/ZnO composite mats enhanced the photocatalytic degradation and the mineralization of both the organic pollutants studied, MB and BPA. In fact, the Au NPs were homogeneously distributed on the surface of the ZnO NPs without the formation of aggregates, thus allowing an effective metal–semiconductor interface, which is fundamental for the improvement of the photocatalytic performance of the ZnO under UV light irradiation. Although the BPA degradation was less effective than that obtained for MB, due to the intrinsic nature of the BPA molecules, the presence of Au significantly improved the photodegradation performance of ZnO for such a persistent pollutant. Therefore, the proposed fabrication approach can be considered a valid alternative to the conventional routes for the development of efficient supported photocatalysts for water remediation applications.

Supplementary Materials: The following are available online at http://www.mdpi.com/2073-4441/11/9/1787/s1, Details on the Debye–Scherrer equation and Kubelka–Munk method, Figure S1: Photos of (**a**) PMMA/ZnO, (**b**) PMMA/ZnO-Au1, (**c**) PMMA/ZnO-Au3, and (**d**) PMMA/ZnO-Au6. Figure S2: SEM images and size distribution analysis of the diameter of the fibers of (**a,b**) PMMA and (**c,d**) PMMA/Zn(CH$_3$CO$_2$)$_2$ mats. Figure S3: (**a**) HRSEM image of PMMA/ZnO with cross-sectional detail. (**b**) Diameter size distribution of the PMMA/ZnO composite mat. Figure S4: TEM images of (**a**) the smaller ZnO NPs formed in the bulk of the polymeric fibers and of (**b**) the ZnO/Au hybrid structure in the PMMA/ZnO-Au1 composite mat. Figure S5: Dark field TEM image of (**a**) the ZnO-Au hybrid structure and EDS mapping of (**b**) Au, (**c**) Zn, and (**d**) O. Figure S6: Kubelka–Munk plots of the composite mats. The energy band gap is extrapolated from a linear regression. Table S1: Assignment of the Raman modes of the PMMA. Figure S7: Raman spectra of PMMA/ZnO-Au3 and PMMA/ZnO-Au6. Figure S8: Evolution of the normalized concentration of (**a**) MB and (**b**) BPA solutions in presence of the developed mats in dark. Table S2: Degradation and mineralization values obtained from the photocatalytic degradation of the MB and BPA aqueous solution in presence of the mats after 20 h under UV light irradiation. Figure S9: The pseudo-first-order reaction kinetics for (**a**) MB and (**b**) BPA, applied on the experimental data obtained in the first 5 h of reaction. Table S3: Photo-degradation rate constants and linear regression coefficients obtained from the linear fitting of the experimental data by using the pseudo-first order model. Figure S10: Photocatalytic degradation activity of (**a**) MB and (**b**) BPA for three consecutive UV irradiation cycles using PMMA/ZnO composite mats.

Author Contributions: L.C. and D.F. conceived and designed the experiments. L.C. realized the materials, performed the characterizations and the photocatalytic experiments, analyzed the data, and accomplished the original draft. S.L. acquired HRSEM images and contributed to the draft editing. A.A. provided the funding and the resources. D.F. supervised the research, validated the results, and edited and reviewed the draft.

Funding: This research received no external funding.

Acknowledgments: The authors kindly acknowledge Alice Scarpellini for the acquisition of the dark-field TEM images and Filippo Drago for the ICP-OES analysis. The useful discussion with Davide Morselli concerning the preparation and the characterization of the materials is also gratefully acknowledged.

Conflicts of Interest: The authors declare no conflict of interest. The funders had no role in the design of the study; in the collection, analyses, or interpretation of data; in the writing of the manuscript, or in the decision to publish the results.

References

1. Hernández-Chover, V.; Bellver-Domingo, Á.; Hernández-Sancho, F. Efficiency of wastewater treatment facilities: The influence of scale economies. *J. Environ. Manage.* **2018**, *228*, 77–84. [CrossRef] [PubMed]
2. Crini, G.; Lichtfouse, E. Advantages and disadvantages of techniques used for wastewater treatment. *Environ. Chem. Lett.* **2019**, *17*, 145–155. [CrossRef]
3. Luo, J.; Zhang, Q.; Cao, J.; Fang, F.; Feng, Q. Importance of monitor and control on new-emerging pollutants in conventional wastewater treatment plants. *J. Geosci. Environ. Prot.* **2018**, *6*, 55–58. [CrossRef]
4. Matamoros, V.; Rodríguez, Y.; Albaigés, J. A comparative assessment of intensive and extensive wastewater treatment technologies for removing emerging contaminants in small communities. *Water Res.* **2016**, *88*, 777–785. [CrossRef] [PubMed]
5. Fiorenza, R.; Bellardita, M.; D'Urso, L.; Compagnini, G.; Palmisano, L.; Scirè, S. Au/TiO$_2$-CeO$_2$ catalysts for photocatalytic water splitting and VOCs oxidation reactions. *Catalysts* **2016**, *6*, 121. [CrossRef]
6. Miklos, D.B.; Remy, C.; Jekel, M.; Linden, K.G.; Drewes, J.E.; Hübner, U. Evaluation of advanced oxidation processes for water and wastewater treatment—A critical review. *Water Res.* **2018**, *139*, 118–131. [CrossRef] [PubMed]
7. Philippe, K.K.; Timmers, R.; Van Grieken, R.; Marugan, J. Photocatalytic disinfection and removal of emerging pollutants from effluents of biological wastewater treatments, using a newly developed large-scale solar simulator. *Ind. Eng. Chem. Res.* **2016**, *55*, 2952–2958. [CrossRef]
8. Amor, C.; Marchão, L.; Lucas, M.S.; Peres, J.A. Application of advanced oxidation processes for the treatment of recalcitrant agro-industrial wastewater: A review. *Water* **2019**, *11*, 205. [CrossRef]
9. Ognibene, G.; Cristaldi, D.A.; Fiorenza, R.; Blanco, I.; Cicala, G.; Scirè, S.; Fragalà, M.E. Photoactivity of hierarchically nanostructured ZnO-PES fibre mats for water treatments. *RSC Adv.* **2016**, *6*, 42778–42785. [CrossRef]
10. Maučec, D.; Šuligoj, A.; Ristić, A.; Dražić, G.; Pintar, A.; Tušar, N.N. Titania versus zinc oxide nanoparticles on mesoporous silica supports as photocatalysts for removal of dyes from wastewater at neutral pH. *Catal. Today* **2018**, *310*, 32–41. [CrossRef]
11. Neghi, N.; Kumar, M.; Burkhalov, D. Synthesis and application of stable, reusable TiO$_2$ polymeric composites for photocatalytic removal of metronidazole: Removal kinetics and density functional analysis. *Chem. Eng. J.* **2019**, *359*, 963–975. [CrossRef]
12. Podasca, V.E.; Buruiana, T.; Buruiana, E.C. Photocatalytic degradation of Rhodamine B dye by polymeric films containing ZnO, Ag nanoparticles and polypyrrole. *J. Photochem. Photobiol. A Chem.* **2019**, *371*, 188–195. [CrossRef]
13. Morselli, D.; Campagnolo, L.; Prato, M.; Papadopoulou, E.L.; Scarpellini, A.; Athanassiou, A.; Fragouli, D. Ceria/gold nanoparticles in situ synthesized on polymeric membranes with enhanced photocatalytic and radical scavenging activity. *ACS Appl. Nano Mater.* **2018**, *1*, 5601–5611. [CrossRef]

14. Colmenares, J.C.; Kuna, E. Photoactive hybrid catalysts based on natural and synthetic polymers: A comparative overview. *Molecules* **2017**, *22*, 790. [CrossRef] [PubMed]

15. Di Mauro, A.; Cantarella, M.; Nicotra, G.; Pellegrino, G.; Gulino, A.; Brundo, M.V.; Privitera, V.; Impellizzeri, G. Novel synthesis of ZnO/PMMA nanocomposites for photocatalytic applications. *Sci. Rep.* **2017**, *7*, 1–12. [CrossRef]

16. Hegedűs, P.; Szabó-Bárdos, E.; Horváth, O.; Szabó, P.; Horváth, K. Investigation of a TiO$_2$ photocatalyst immobilized with poly(vinyl alcohol). *Catal. Today* **2016**, *285*, 179–186. [CrossRef]

17. Morselli, D.; Valentini, P.; Perotto, G.; Scarpellini, A.; Pompa, P.P.; Athanassiou, A.; Fragouli, D. Thermally-induced in situ growth of ZnO nanoparticles in polymeric fibrous membranes. *Compos. Sci. Technol.* **2017**, *149*, 11–19. [CrossRef]

18. Pinto, J.; Morselli, D.; Bernardo, V.; Notario, B.; Fragouli, D.; Rodriguez-Perez, M.A.; Athanassiou, A. Nanoporous PMMA foams with templated pore size obtained by localized in situ synthesis of nanoparticles and CO$_2$ foaming. *Polymer* **2017**, *124*, 176–185. [CrossRef]

19. Feng, J.; Athanassiou, A.; Bonaccorso, F.; Fragouli, D. Enhanced electrical conductivity of poly(Methyl methacrylate) filled with graphene and in situ synthesized gold nanoparticles. *Nano Futur.* **2018**, *2*, 025003. [CrossRef]

20. Demir, M.M.; Gulgun, M.A.; Menceloglu, Y.Z.; Erman, B.; Abramchuk, S.S.; Makhaeva, E.E.; Khokhlov, A.R.; Matveeva, V.G.; Sulman, M.G. Palladium nanoparticles by electrospinning from Poly(acrylonitrile-co-acrylic acid)-PdCl2 solutions. relations between preparation conditions, particle size, and catalytic activity. *Macromolecules* **2004**, *37*, 1787–1792. [CrossRef]

21. Andrade, G.R.S.; Nascimento, C.C.; Lima, Z.M.; Teixeira-Neto, E.; Costa, L.P.; Gimenez, I.F. Star-shaped ZnO/Ag hybrid nanostructures for enhanced photocatalysis and antibacterial activity. *Appl. Surf. Sci.* **2017**, *399*, 573–582. [CrossRef]

22. Thanh, Q.; Ta, H.; Park, S.; Noh, J. Ag nanowire/ZnO nanobush hybrid structures for improved photocatalytic activity. *J. Colloid Interface Sci.* **2017**, *505*, 437–444. [CrossRef] [PubMed]

23. He, X.; Yang, D.-P.; Zhang, X.; Liu, M.; Kang, Z.; Lin, C.; Jia, N.; Luque, R. Waste eggshell membrane-templated CuO-ZnO nanocomposites with enhanced adsorption, catalysis and antibacterial properties for water purification. *Chem. Eng. J.* **2019**, *369*, 621–633. [CrossRef]

24. Ussia, M.; Di Mauro, A.; Mecca, T.; Cunsolo, F.; Nicotra, G.; Spinella, C.; Cerruti, P.; Impellizzeri, G.; Privitera, V.; Carroccio, S.C. ZnO-pHEMA nanocomposites: An Ecofriendly and Reusable Material for Water Remediation. *ACS Appl. Mater. Interfaces* **2018**, *10*, 40100–40110. [CrossRef] [PubMed]

25. Su, Y.Q.; Zhu, Y.; Yong, D.; Chen, M.; Su, L.; Chen, A.; Wu, Y.; Pan, B.; Tang, Z. Enhanced Exciton Binding Energy of ZnO by Long-Distance Perturbation of Doped Be Atoms. *J. Phys. Chem. Lett.* **2016**, *7*, 1484–1489. [CrossRef] [PubMed]

26. Guidelli, E.J.; Baffa, O.; Clarke, D.R. Enhanced UV Emission from Silver/ZnO and Gold/ZnO Core-Shell Nanoparticles: Photoluminescence, Radioluminescence, and Optically Stimulated Luminescence. *Sci. Rep.* **2015**, *5*, 1–11. [CrossRef]

27. Li, P.; Wei, Z.; Wu, T.; Peng, Q.; Li, Y. Au-ZnO hybrid nanopyramids and their photocatalytic properties. *J. Am. Chem. Soc.* **2011**, *133*, 5660–5663. [CrossRef]

28. He, W.; Kim, H.-K.; Wamer, W.G.; Melka, D.; Callahan, J.H.; Yin, J.-J. Photogenerated charge carriers and reactive oxygen species in ZnO/Au hybrid nanostructures with enhanced photocatalytic and antibacterial activity. *J. Am. Chem. Soc.* **2014**, *136*, 750–757. [CrossRef]

29. Rabanal, M.E. Solvothermal synthesis of Ag/ZnO and Pt/ZnO nanocomposites and comparison of their photocatalytic behaviors on dyes degradation. *Adv. Powder Technol.* **2016**, *27*, 983–993. [CrossRef]

30. Gupta, J.; Mohapatra, J.; Bahadur, D. Visible light driven mesoporous Ag-embedded ZnO nanocomposites: Reactive oxygen species enhanced photocatalysis, bacterial inhibition and photodynamic therapy. *Dalt. Trans.* **2017**, *46*, 685–696. [CrossRef]

31. Putri, N.A.; Fauzia, V.; Iwan, S.; Roza, L.; Umar, A.A.; Budi, S. Mn-doping-induced photocatalytic activity enhancement of ZnO nanorods prepared on glass substrates. *Appl. Surf. Sci.* **2018**, *439*, 285–297. [CrossRef]

32. Kong, J.-Z.; Zhai, H.-F.; Zhang, W.; Wang, S.-S.; Zhao, X.-R.; Li, M.; Li, H.; Li, A.-D.; Wu, D. Visible light-driven photocatalytic performance of N-doped ZnO/g-C 3 N 4 Nanocomposites. *Nanoscale Res. Lett.* **2017**. [CrossRef] [PubMed]

33. Waiskopf, N.; Ben-Shahar, Y.; Banin, U. Photocatalytic hybrid semiconductor–metal nanoparticles; from synergistic properties to emerging applications. *Adv. Mater.* **2018**, *30*, 1–10. [CrossRef] [PubMed]

34. Picciolini, S.; Castagnetti, N.; Vanna, R.; Mehn, D.; Bedoni, M.; Gramatica, F.; Villani, M.; Calestani, D.; Pavesi, M.; Lazzarini, L.; et al. Branched gold nanoparticles on ZnO 3D architecture as biomedical SERS sensors. *RSC Adv.* **2015**, *5*, 93644–93651. [CrossRef]

35. Zhang, W.; Wang, W.; Shi, H.; Liang, Y.; Fu, J.; Zhu, M. Surface plasmon-driven photoelectrochemical water splitting of aligned ZnO nanorod arrays decorated with loading-controllable Au nanoparticles. *Sol. Energy Mater. Sol. Cells* **2018**, *180*, 25–33. [CrossRef]

36. Cheng, Y.; Jiao, W.; Li, Q.; Zhang, Y.; Li, S.; Li, D.; Che, R. Two hybrid Au-ZnO aggregates with different hierarchical structures: A comparable study in photocatalysis. *J. Colloid Interface Sci.* **2017**, *509*, 58–67. [CrossRef]

37. Muravitskaya, A.; Rumyantseva, A.; Kostcheev, S.; Dzhagan, V.; Stroyuk, O.; Adam, P.-M. Enhanced raman scattering of ZnO nanocrystals in the vicinity of gold and silver nanostructured surfaces. *Opt. Express* **2016**, *24*, A168–A173. [CrossRef]

38. Xie, W.; Li, Y.; Sun, W.; Huang, J.; Xie, H.; Zhao, X. Surface modification of ZnO with Ag improves its photocatalytic efficiency and photostability. *J. Photochem. Photobiol. A Chem.* **2010**, *216*, 149–155. [CrossRef]

39. Akira, M.; Yanzhi, R.; Kimihiro, M.; Hiroshi, I.; Yukio, M.; Isao, N.; Yukihiro, O. Two-dimensional fourier-transform raman and near-infrared correlation spectroscopy studies of poly(methyl methacrylate) blends: 1. Immiscible blends of poly(methyl methacrylate) and atactic polystyrene. *Vib. Spectrosc.* **2000**, *24*, 171–180.

40. Rumyantseva, A.; Kostcheev, S.; Adam, P.M.; Gaponenko, S.V.; Vaschenko, S.V.; Kulakovich, O.S.; Ramanenka, A.A.; Guzatov, D.V.; Korbutyak, D.; Dzhagan, V.; et al. Nonresonant surface-enhanced raman scattering of ZnO quantum dots with Au and Ag nanoparticles. *ACS Nano* **2013**, *7*, 3420–3426. [CrossRef]

41. Antony, A.; Poornesh, P.; Kityk, I.V.; Ozga, K.; Jedryka, J.; Philip, R.; Sanjeev, G.; Petwal, V.C.; Verma, V.P.; Dwivedi, J. Methodical engineering of defects in Mn X Zn 1-X O(x = 0.03, and 0.05) nanostructures by electron beam for nonlinear optical applications: A new insight. *Ceram. Int.* **2019**, *45*, 8988–8999. [CrossRef]

42. Milekhin, A.G.; Sveshnikova, L.L.; Duda, T.A.; Yeryukov, N.A.; Rodyakina, E.E.; Gutakovskii, A.K.; Batsanov, S.A.; Latyshev, A.V.; Zahn, D.R.T. Surface-enhanced Raman spectroscopy of semiconductor nanostructures. *Phys. E Low Dimensional Syst. Nanostruct.* **2016**, *75*, 210–222. [CrossRef]

43. Barrios-Estrada, C.; de Jesús Rostro-Alanis, M.; Muñoz-Gutiérrez, B.D.; Iqbal, H.M.N.; Kannan, S.; Parra-Saldívar, R. Emergent contaminants: Endocrine disruptors and their laccase-assisted degradation—A review. *Sci. Total Environ.* **2018**, *612*, 1516–1531. [CrossRef] [PubMed]

44. Bilic, A.; Reimers, J.R.; Hush, N.S.; Hoft, R.C.; Ford, M.J.; Biosciences, M. Adsorption of benzene on copper, silver, and gold surfaces. *Adsorpt. J. Int. Adsorpt. Soc.* **2006**, *2*, 1093–1105.

45. Ide, Y.; Matsuoka, M.; Ogawa, M. Efficient visible-light-induced photocatalytic activity on gold-nanoparticle-supported layered titanate. *J. Am. Chem. Soc.* **2010**, *132*, 16762–16764. [CrossRef]

46. Kong, L.; Jiang, Z.; Lai, H.H.; Xiao, T.; Edwards, P.P. Progress in natural science: Materials international does noble metal modi fi cation improve the photocatalytic activity of BiOCl? *Prog. Nat. Sci. Mater. Int.* **2013**, *23*, 286–293. [CrossRef]

47. Liu, H.; Hu, Y.; Zhang, Z.; Liu, X.; Jia, H.; Xu, B. Synthesis of spherical Ag/ZnO heterostructural composites with excellent photocatalytic activity under visible light and UV irradiation. *Appl. Surf. Sci.* **2015**, *355*, 644–652. [CrossRef]

48. Gomes, F.; Lopes, A.; Bednarczyk, K.; Gmurek, M.; Stelmachowski, M.; Id, A.Z.; Quinta-Ferreira, M.E.; Costa, R.; Quinta-ferreira, R.M.; Martins, R.C. Effect of noble metals (Ag, Pd, Pt) Loading over the efficiency of TiO$_2$ during photocatalytic ozonation on the toxicity of parabens. *ChemEngineering* **2018**, *2*, 4. [CrossRef]

49. Erjavec, B.; Hudoklin, P.; Perc, K.; Tišler, T.; Dolenc, M.S.; Pintar, A. Glass fiber-supported TiO$_2$ photocatalyst: Efficient mineralization and removal of toxicity/estrogenicity of bisphenol A and its analogs. *Appl. Catal. B Environ.* **2016**, *183*, 149–158. [CrossRef]
50. Lee, K.M.; Lai, C.W.; Ngai, K.S.; Juan, J.C. Recent developments of zinc oxide based photocatalyst in water treatment technology: A review. *Water Res.* **2016**, *88*, 428–448. [CrossRef]

Article

Preparation of Biomass Activated Carbon Supported Nanoscale Zero-Valent Iron (Nzvi) and Its Application in Decolorization of Methyl Orange from Aqueous Solution

Bo Zhang * and Daping Wang

School of Metallurgical and Material Engineering, Hunan University of Technology, Taishan Road 88, Zhuzhou 412007, China
* Correspondence: 13747@hut.edu.cn

Received: 12 July 2019; Accepted: 8 August 2019; Published: 12 August 2019

Abstract: The nanoscale zero-valent iron (nZVI) has great potential to degrade organic polluted wastewater. In this study, the nZVI particles were obtained by the pulse electrodeposition and were loaded on the biomass activated carbon (BC) for synthesizing the composite material of BC-nZVI. The composite material was characterized by SEM-EDS and XRD and was also used for the decolorization of methyl orange (MO) test. The results showed that the 97.94% removal percentage demonstrated its promise in the remediation of dye wastewater for 60 min. The rate of MO matched well with the pseudo-second-order model, and the rate-limiting step may be a chemical sorption between the MO and BC-nZVI. The removal percentage of MO can be effectively improved with higher temperature, larger BC-nZVI dosage, and lower initial concentration of MO at the pH of 7 condition.

Keywords: biomass activated carbon; methyl orange; pulse electrodeposition; zero valent iron nanoparticles

---

1. Introduction

Dye has become a widespread environmental pollution problem because of its wide application in industry such as paper, textiles, plastic, and leather tanning industries in the past several decades [1,2]. Over 100,000 commercial dyes are associated with an annual production rate of over 800,000 tons [3]. Large amounts of dye-containing effluents pose great challenges to the environment because of their strong color, complex structure, stability, and low biodegradability [4]. Therefore, it is of vital importance to remove dyes from wastewater to protect the aquatic life and alleviate the crisis of water pollution.

Many methods such as adsorption [5], reduction [6], advanced oxidation processes [7], coagulation [8], membrane separation [9], and biological methods [10] are used to remove dyes from wastewater. In recent years, nanoscale zero-valent iron (nZVI) particle has been extensively used as a new tool for the treatment of wastewater contaminated with various pollutants as a result of its small particle size, large specific surface area, and high reactivity [11]. Unfortunately, there are still some technical challenges in the use of nZVI. On the one hand, because of interparticle Van der Waals and magnetic interactions, nZVI particles are prone to agglomeration, resulting in the significant decrease of their dispersibility [12]. On the other hand, nanoparticles tend to be oxidized, and the formation of oxide layers easily block the serviceable active surface sites, which finally diminishes the reactivity [13]. To get rid of these shortcomings, supporting nZVI particles is an option. For the past few years, most of previous studies have loaded nZVI particles on some porous materials as a carrier, such as kaolinite [14], bentonite [15], resin [16], activated carbon [17], mesoporous carbon [18], mesoporous silica [19], titanium oxide [20], pumice [1,21], and grapheme [22], through the liquid phase reduction method. Nevertheless, there still has some problems associated with method, such as low production efficiency, high production cost, and large amounts of hydrogen the generation during the preparation process [23,24]. All of these problems lead to the limitations of large-scale application.

In this study, the compound material was prepared with two steps. First, the nZVI particles were directly obtained from steel scrap using the pulse electrodeposition. Second, the nZVI particles were quickly supported to the biomass activated carbon (BC) under the mechanical agitation condition. The obtained BC-nZVI material was characterized and its decontamination abilities were tested by the removal of methyl orange (MO). In addition, the effects of dosage, initial pH, initial concentration, and temperature on the removal percentage of MO were investigated in conjunction with the analyses of mechanism and kinetics. It is hoped that this study can provide a new preparation technique for nZVI composites for wastewater treatment.

2. Materials and Methods

2.1. Materials and Chemicals

The biomass activated carbon was carbonized from coconut shell with particle size of 2~4 mm, filling density of 0.5~0.55 g/mL, PH value of 6.5~7.5, and iodine adsorption value of 850~1000 mg/g. This was supplied by Lu-yuan Co. Ltd., Mianyang, China. Ferrous sulfate septihydrate (FeSO$_4$·7H$_2$O, ≥98.5%), methyl orange (MO, 99%), sodium hydroxide (NaOH, 98%), hydrochloric acid (HCl, 36%), sodium dodecyl benzene sulfonate (DBS, 99%), thiourea (≥98%), and absolute ethanol (99%) were obtained from Sinopharm Chemical Reagent Co. Ltd., Shanghai, China. All the chemicals were analytical reagent grade and deionized water from a Liceng UPA-L system (18.2 MΩ/cm, 25 °C) was used for all experiments.

2.2. Preparation of nZVI and Composite with Biomass Activated Carbon

FeSO$_4$·7H$_2$O was dissolved in deionized water to prepare an electrolyte with Fe^{2+} of 30 g/L. The additive (thiourea, 0.5 g/L and DBS, 1.0 g/L) was dispersed and emulsified in deionized water in an ultrasonic cleaner (VGT SONIC-L30 300 w, 28 kHz) and added to the electrolyte. The electrolysis was carried out at 50~60 °C and the anode plate for electrolysis was a waste carbon steel (Q235) plate, which was used after surface treated. The cathode plate for electrolysis was a stainless-steel box and several ultrasonic vibrators were added in the box. The ultrasonic power was 28 KHz. The cathode and the anode were separated by a filter membrane and the structure of the electrolytic cell is shown in Figure 1. After 10 h of electrolysis, the electrolyte near the cathode plate was rapidly pumped out and filtration, and the leached residue was washed with deionized water and absolute ethanol for three times, respectively. Subsequently, DBS and absolute ethanol were intermixed with leached residue and dispersed in a mechanical disperser (FS400D, 2000 rpm, Qiwei instrument Co. Ltd., Hangzhou, China). Finally, an emulsion of nZVI was prepared after 12 h of dispersion under nitrogen protective.

The electrolytic process occurs at the cathode and anode as follows [25]:
Anode:

$$Fe - 2e = Fe^{2+} \tag{1}$$

Cathode:

$$Fe^{2+} + 2e = Fe \tag{2}$$

The iron particles generated on the cathode plate were quickly stripped and emulsified in the electrolyte under the action of cavitation effect of ultrasonic oscillation and combined with dispersant. The polymer molecular chain of dispersant was negatively charged, which can form a brush surface layer to hinder the aggregation. Under the action of dispersant, iron particles were difficult to agglomerate and oxidize.

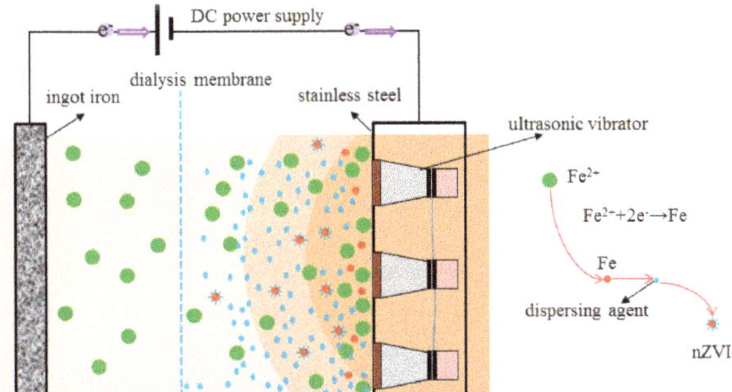

Figure 1. Electrolytic cell for preparation of nanoscale zero-valent iron (nZVI).

The emulsion of nZVI was transferred to a conical flask (1000 mL) for 100 mL, and 200 mL deionized water and 200 g biomass activated carbon were added into conical flask. Then, the conical flask was sealed with a sealing plug and shocked on a horizontal vibrator (HY-5A) with a frequency of 240 rpm. After 10 h of shocking, the mixture was cleaned three times by absolute ethanol to remove the excess emulsion. Finally, the freshly composite material BC-nZVI was dried at 30 °C in the vacuum drying oven for 12 h and kept in a nitrogen atmosphere prior to use.

2.3. Characterization and Analytic Methods

The surface morphology images of BC-nZVI were obtained with a scanning electron microscope (SEM) (SIGMA 300; Carl Zeiss AG, Oberkochen, Hallbergmoos, Germany) operating at 30 kV. The crystal structure and composition phase was analyzed by X-ray diffraction (XRD) (Philips-X'Pert Pro MPD, Malvern Panalytical, Almelo, Holland, The Netherlands). The sample of BC-nZVI was dissolved by HCl and the total iron ions in acid solution were analyzed by inductively Coupled plasma spectrometer (ICP-MS 8000; Perkinelmer, Waltham, MA, USA) to determine the actual load of nZVI on BC. The concentration of MO solution was measured using a UV-Spectrophotometer (760CRT, Shanghai precision scientific instrument co., LTD, Shanghai, China) at λ_{max} = 464 nm [26].

2.4. Batch Experiments

The removal percentage of MO by BC-nZVI was evaluated by batch experiments. The degradation test for MO was carried out in a 250 mL conical flask and the volume of the experimental solution was 150 mL. The flask was placed in the horizontal vibrator and wobbled at 240 rpm after the BC-nZVI was added. The initial pH value of solution was adjusted by 0.01 mol/L HCl and 0.01 mol/L NaOH. The five major factors (MO initial concentration (20–200 mg/L), weight of BC-nZVI dosage (0.25–0.75 mg/L), reaction temperature (20°C~80°C), pH (3.0~10.0), and interaction time (60–120 min)) were considered for optimizing MO degradation by BC-nZVI. The samples (4 mL) were collected within a specified time and the concentration of MO was estimated using UV-Spectrophotometer after centrifuged. In order to ensure the quality of the data, all experiments were conducted in three copies and the average value was reported.

The MO removal percentage was calculated by Equation (3):

$$R(\%) = \frac{(C_0 - C_e) \times 100}{C_0} \tag{3}$$

The equilibrium removal capacity of MO (q_e (mg/g)) was calculated by Equation (4):

$$q_e = \frac{(C_0 - C_e) \times V}{W} \tag{4}$$

where C_0 (mg/L) is initial concentration of MO, C_e (mg/L) is the equilibrium concentrations of MO, V (L) is the experimental solution volume, and W (g) is dry weight of nZVI in BC-nZVI used.

3. Results and Discussions

3.1. Characterization of nZVI and BC-nZVI

The morphologies of nZVI and BC-nZVI were determined using SEM and presented in Figure 2. Figure 2a shows that the nZVI nanoparticles were substantially nearly smooth and spherical with sizes ranging from 40 to 80 nm. The nanoparticles were slightly agglomerated, and this may have occurred during the detection process when the emulsion of nZVI was placed on the observation table and rapidly dried by washing ear ball. The surface of BC presents a porous structure, which provides a good platform for loading the nZVI as shown in Figure 2b. The nanoparticles were substantially evenly distributed on the biomass activated carbon surface and pores under mechanical loading condition.

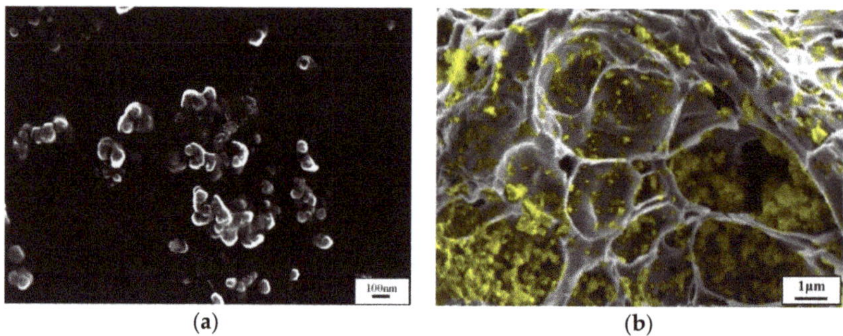

(a)   (b)

Figure 2. SEM images of (**a**) nZVI, (**b**) biomass activated carbon (BC)-nZVI.

The XRD patterns of nZVI, BC, and BC-nZVI were obtained and presented in Figure 3. No characteristic diffraction peaks of Fe were observed because of its weak crystallization and the inclusion of DBS. The characteristic peaks have several small peaks at $2\theta = 17.41°$, $19.24°$, and $3.84°$ on nZVI wave lines, which represent iron oxide [22], indicating a certain oxidation in the nanoparticles. In the characteristic peak of the BC-nZVI wave lines, there were obvious carbon peaks at $2\theta = 26.60°$, $43.45°$, $54.79°$ [23] and the iron oxide peak disappears, indicating that the loading was beneficial to prevent the oxidation of the nZVI.

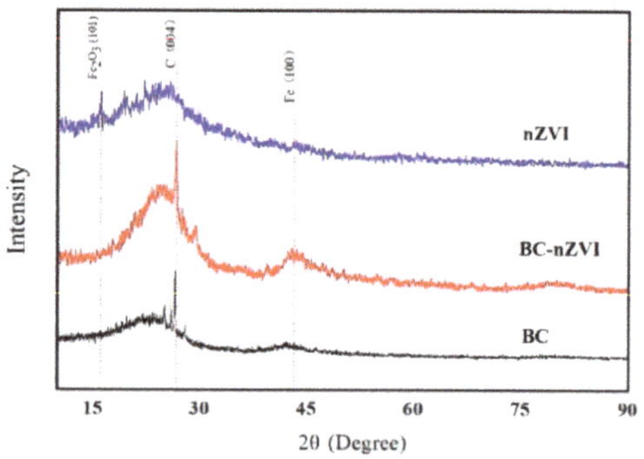

Figure 3. XRD pattern of nZVI, BC, and BC-nZVI.

3.2. Removal Efficiencies of MO

The removal percentage of MO was investigated using BC, nZVI, and BC-nZVI, respectively, as shown in Figure 4. Obviously, the total removal percentage of BC-nZVI and nZVI are significantly higher than that of BC as a result of the contribution of nZVI. But for nZVI and BC-nZVI, the different tendencies were presented. In the early stage (within 20 min), nZVI holds the higher removal percentage than that of BC-nZVI. The reason behind this fact is that the nZVI was in direct contact with the aqueous solution and the reaction was fast, while the removal percentage of BC-nZVI is lower because the pore structure of BC prevents the contact bewteen nZVI and MO. However, the opposite results can be found after 20 min, and the BC-nZVI shows the higher removal percentage than that of nZVI. This is due to the fact that the sole nZVI particle was easily oxidized than BC-nZVI composite material, resulting in the decreased reaction activity.

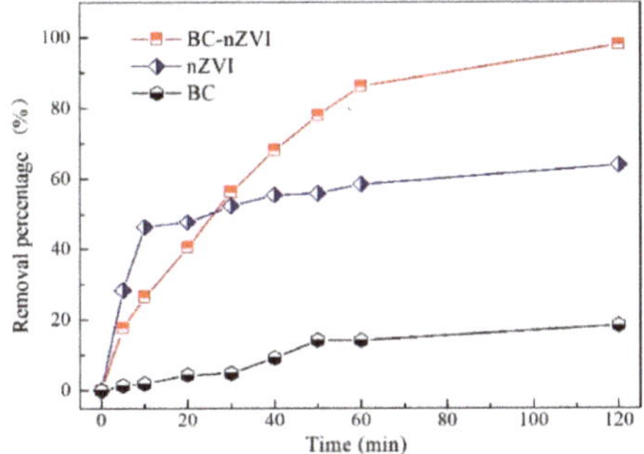

Figure 4. Comparative degradation of MO using different materials. The dosage of BC, nZVI, and BC-nZVI were 3 g/L, 0.5 g/L, and 0.5 g/L (the dry weight of nZVI in BC-nZVI used) with an initial MO concentration of 200 mg/L, temperature of 25 °C, and original pH.

Figure 5 showed the UV–vis absorption spectra of the MO before and after the addition of BC-nZVI. For the original MO, there were two main absorption bands at 464 and 270 nm, which were attributed to a conjugate structure formed by the azo band under the effects of the electronic-donation of dimethylamino group and the π–π * transition of aromatic rings [27]. Obviously, the band intensity of 464 and 270 nm decreased with reaction time, and almost disappeared after reaction 50 min. It proves the degradation of MO through the cracking of azo bonds. The XRD pattern of BC-nZVI after interaction with MO was represented in Figure 6. The characteristic peaks of 27.87°, 50.17° present the iron (III) oxide Fe_2O_3. With the increase of reaction time, the intensity of Fe_2O_3 characteristic peaks increased. The radicals of H were generated by the reaction nZVI nanoparticles and H_2O or hydrogen ion [28,29], which caused the azo bond to open and consequently the absorption bands at 464 nm and 270 nm vanished.

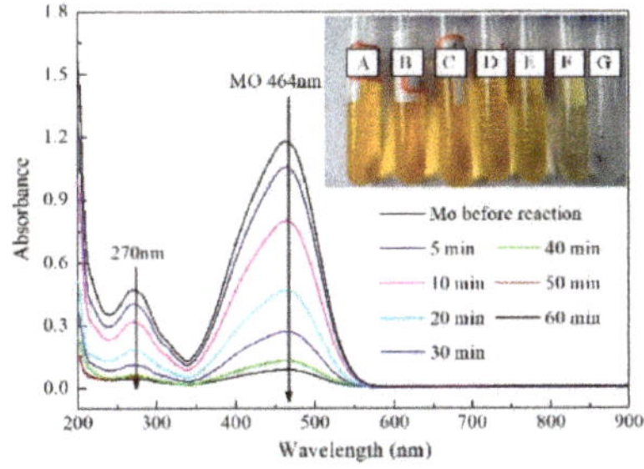

Figure 5. UV–vis patterns of degradation of MO.

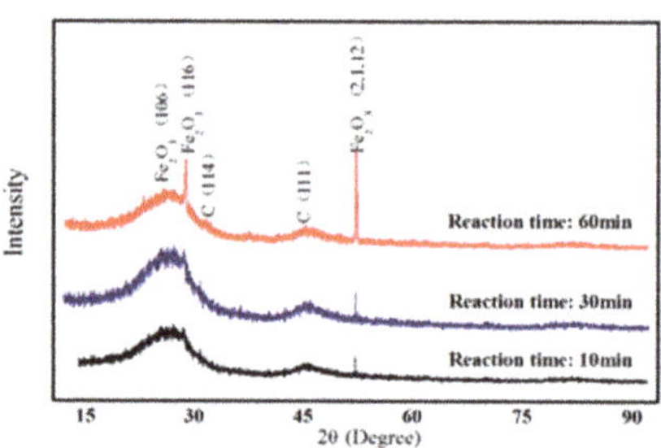

Figure 6. XRD pattern of BC-nZVI after reaction.

The maximum removal capacities of BC-nZVI in this study were compared with those of other absorbents prepared by liquid phase reduction method, as shown in Table 1. Clearly, the present adsorbent (BC-nZVI) shows the same level of removal capacity as other efficient absorbents, such as B-nZVI, HJ-NZVI, nZVI/BC, and Bio-nZVI. However, it should be emphasized that the BC-nZVI in this study was synthesized by pulse electrodeposition and mechanical agitation method, and this method is easier and more cost-effective than the liquid phase reduction method.

Table 1. Comparison of MO removal with different absorbents reported in literature.

Absorbent	Dosage	Initial pH	Initial Concentration	T °C	Reaction Time	Removal Capacity	Refs.
B-nZVI	0.5 g/L	Or	200 mg/L	30	60 min	93.75%	Chen et al. 2011 [13]
HJ-NZVI5	1.0 g/L	Or	200 mg/L	25	45 min	72.8%	Li et al. 2017 [30]
nZVI/BC	0.6 g/L	Or	200 mg/L	NA	60 min	96.17%	Yang et al. 2018 [31]
B */nZVI/Pd	0.5 g/L	Or	200 mg/L	27	90 min	91.87%	Wang et al. 2013 [32]
nZVI/Pd	0.5 g/L	Or	200 mg/L	27	90 min	85%	Wang et al. 2013 [32]
BC-nZVI	0.5 g/L	Or	200 mg/L	25	120 min	97.94%	This study
nZVI	0.5 g/L	Or	200 mg/L	25	120 min	63.9%	This study

B: synthesized bentonite; HJ: Hangjin clay; BC: biochar; Bio: Biogenic; B *: Functional clay; Or is 6.17; NA: not available.

4. Impact of Operational Parameters

4.1. Effect of Dosage

As illustrated in Figure 7a, the decolorization kinetics of MO was dependent on the BC-nZVI dosage. The larger the BC-nZVI dosage, the higher the removal percentage. The concentration of MO decreases dramatically in the initial 20 min at the highest BC-nZVI dosage of 0.75 g/L, while the removal percentage at 0.25 g/L was sluggish. Moreover, for the dosages of 0.25, 0.5, and 0.75 g/L, the maximum removal percentages of MO were 68.96%, 94.41%, and 97.43%, and the removal capacities of BC-nZVI were 344.1 mg/g, 300.8 mg/g, and 181.2 mg/g, respectively. Due to the loading effect of BC, the oxidation of nZVI was slowed down, resulting in the highest removal percentage.

When the amount of BC-nZVI dosage was insufficient, the slower kinetics and lower removal percentage were observed for MO removal, but the removal capacity of BC-nZVI was higher [32]. At a sufficient amount of BC-nZVI, enough nZVI activity sites was provided to react with MO in the initial stage, so the reaction kinetics were faster, but the removal capacity of BC-nZVI was lower.

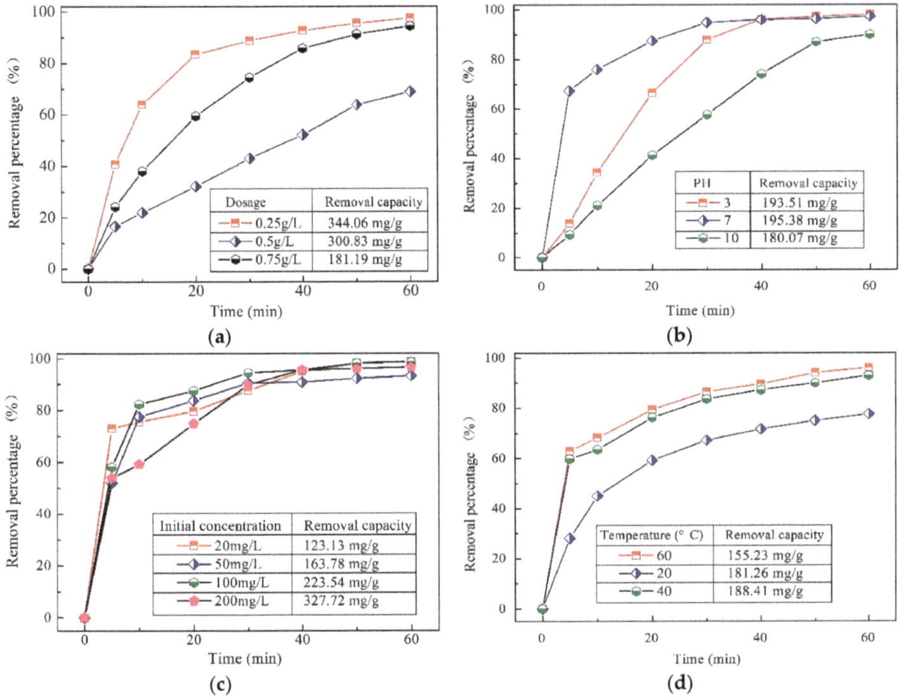

Figure 7. (**a**) Effect of dosage on degradation of MO, with an initial MO concentration of 200 mg/L, temperature of 25 °C, and original pH; (**b**) Effect of the pH values on degradation of MO, in the context of 0.5 g/L BC-nZVI with an initial MO concentration of 50 mg/L; (**c**) Effect of initial concentration on degradation of MO, at the original pH using 0.5 g/L BC-nZVI; (**d**) Effect of temperature on degradation of MO, in the context of 0.5 g/L BC-nZVI with an initial MO concentration of 50 mg/L, temperature of 25 °C, and the original pH.

4.2. Effect of pH

The effects of initial pH on MO removal were illustrated in Figure 7b. With the increase of pH from 3.0 to 10.0, the removal percentage of MO was declined from 97.75% to 89.13%, and decreased from 76.01% to 21.08% in the first 10 min. It suggests that the lower pH value was beneficial to MO reduction by BC-nZVI. The possible explanation is that the more H^+ was released at lower pH values, which could accelerate the corrosion of nZVI particles, and eliminate ferrous hydroxide on the surface of nZVI particles to generate fresh active sites. The removal capacities of BC-nZVI were 193.5 mg/g, 195.4 mg/g, and 180.1 mg/g as the pH were 3.0, 7.0, and 10.0, respectively. The alkaline pH reduced the removal capacity of BC-nZVI because OH^- would significantly enhance the formation of the iron hydroxide, which formed a surface layer on the nZVI particles and inhibited further reactions [20].

4.3. Effect of Initial Concentration

The effects of initial MO concentrations ranging from 20 to 200 mg/L on removal percentage was shown in Figure 7c. The maximum decolorization efficiencies were 98.8%, 98.7%, 96.7%, and 93.3%, and the decolorization efficiencies in first 20 min were 77.6%, 82.3%, 75.6%, and 59.2% at initial concentrations of 20 mg/L, 50 mg/L, 100 mg/L, and 200 mg/L, respectively. The removal percentage with the initial concentration of 200 mg/L was significantly lower than that of the other lower initial concentration. It may be caused by that the highly initial MO concentration leads to a competition effect among the MO molecules and a decline in decolorization efficiency [33,34]. However, with the increase of initial concentration, the removal capacity of nZVI was obviously improved. At a higher initial concentration, the removal capacity of nZVI was significant higher, at 327.7 mg/g.

4.4. Effect of Temperature

As shown in Figure 7d, the removal percentage of MO increased with increasing temperature. The final removal percentages of MO were 77.6%, 93.1%, and 96.2%, respectively, for 20, 40, and 60 °C. Similar results were reported previously [35–37]. There may be two possible explanations for this phenomenon: (1) The mobility of MO was increased by higher temperature, and (2) the activation energy of decolorizing reaction was increased as the temperature rose.

4.5. Kinetics of Degradation of MO

In order to investigate the mechanism of degradation, the pseudo-first-order kinetics model (PFO) (Equation (5)) and the pseudo-second-order model (PSO) (Equation (6)) were generally used to test the degradation of MO using BC-nZVI, which can be expressed as the following equation [13,29]:

$$\ln(q_e - q_t) = \ln q_e - k_1 t \tag{5}$$

$$t/q_1 = 1/(k_2 q_e^2) + t/q_e \tag{6}$$

where q_e and q_t (mg/g) are the amount of MO adsorbed per gram of BC-nZVI at equilibrium and at time t, respectively, while k_1 (min^{-1}) and k_2 (g/(mg·min)) are the PFO and PSO rate constants, respectively.

The effect of initial MO concentrations on the removal efficiency was calculated with these two kinetic models, and the fitting of the kinetics data was shown in Figure 8 and Table 2.

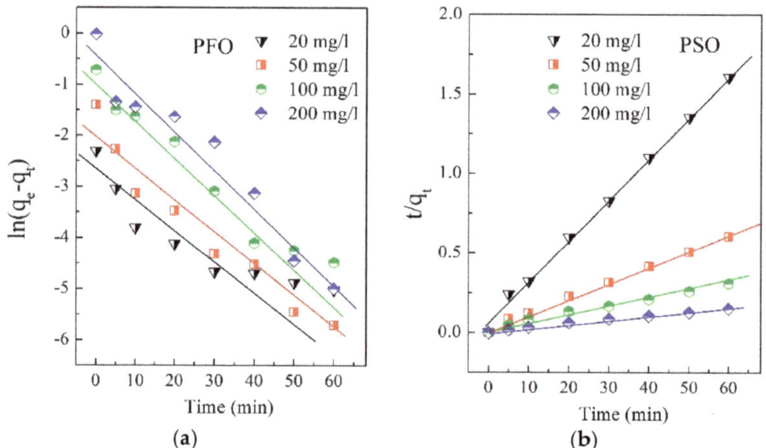

Figure 8. Kinetic modeling of MO adsorption using pseudo-first-order (PFO) (**a**) and pseudo-secondo-rder (PSO) model (**b**).

Table 2. Kinetic parameters for MO removal.

C$_0$ (mg/L)	q$_e$ (mg/g)	PFO			PSO		
		q$_{e}$, cal	k$_1$	R^2	q$_{e}$, cal	K$_2$	R^2
20	123.13	121.93	0.0749	0.8231	124.29	0.0259	0.997
50	163.78	161.77	0.0647	0.9484	163.50	0.0098	0.9974
100	223.54	211.06	0.067	0.9597	229.42	0.0048	0.9879
200	327.72	319.11	0.0401	0.9466	329.98	0.0025	0.9933

Significantly, the pseudo-second-order model matched better with the data (R^2 = 0.988) than the pseudo-first-order model (R^2 = 0.823), suggesting that the chemical sorption may be the main rate-limiting step between the MO and BC-nZVI.

The rate constants for a pseudo second-order reaction were 0.0259 g/(mg·min) for 20 mg/L, 0.0048 g/(mg·min) for 100 mg/L, and 0.0025 g/(mg·min) for 200 mg/L, respectively. The result indicates that the degradation of MO occurs in the interface of BC-nZVI [20,22], hence the rate of degradation was closely linked to the initial concentration of MO and the active surface sites of BC-nZVI, as discussed in the previous section. It should be emphasized that the BC-nZVI materials in our study were prepared by the mechanical agitation method and most of nZVI particles are mainly distributed in the outer layer of BC. It means the MO could easily contact with the nZVI particles, so the chemical sorption is the main rate-limiting step compared to the mass transfer process. However, some other X-nZVI materials synthesized by liquid phase reduction method agreed with the pseudo-first-order model [14–18]. Here, X presents the carrier, such as clay [15], activated carbon [17], and mesoporous carbon [18]. This is due to that the nZVI particles can be uniform distributed in both the outer layer and internal space of carriers (X) using the liquid phase reduction method. Therefore, the mass transfer may be the main limiting step under this condition.

5. Conclusions

The present study demonstrated that BC-nZVI particles can be used to efficiently degrade MO in aqueous solution by cleaving the azo linkages. SEM, XRD, UV–vis, and batch experiments confirmed the characteristics of this composite: (1) The nZVI nanoparticles prepare dby means of pulse electrodeposition were substantially nearly smooth and spherical with sizes ranging from 40 to 80 nm, and successfully loaded on the surface and inside the pores of BC; (2) BC as a dispersant and stabilizer, which reduced the extent of aggregation of nZVI and therefore enhanced reactivity; (3) compared with other adsorbents prepared by liquid phase reduction method, BC-nZVI reflected with the same maximum removal rate; and (4) batch experiments show that various parameters such as dosage, pH, initial concentration of MO, and temperature did affect the removal of MO. In addition, the BC-nZVI in this study was synthesized by pulse electrodeposition and mechanical agitation method, and this method is easier and more cost-effective than the liquid phase reduction method, and the 97.94% removal percentage demonstrated its promise in the remediation of dye wastewater.

Author Contributions: B.Z. designed and carried out this research. D.W. prepared and analyzed the data. B.Z. wrote this paper. All authors have approved the manuscript.

Funding: This study was funded by the Specialized Research Fund for the National Natural Science Foundation of China (No. 51504090), and the Natural Science Foundation of Hunan Province, China (No. 2019JJ60062).

Acknowledgments: The authors are grateful to the three anonymous reviewers for their insightful and constructive comments, which greatly improved the quality of the paper.

Conflicts of Interest: The authors declare no conflict of interest.

References

1. Dong, H.; Deng, J.; Xie, Y.; Zhang, C.; Jiang, Z.; Cheng, Y.; Hou, K.; Zeng, G. Stabilization of nanoscale zero-valent iron (nZVI) with modified biochar for Cr(VI) removal from aqueous solution. *J. Hazard. Mater.* **2017**, *332*, 79–86. [CrossRef] [PubMed]
2. Bernardi, F.; Fecher, G.; Alves, M.; Morais, J. Unraveling the Formation of Core–Shell Structures in Nanoparticles by S-XPS. *J. Phys. Chem. Lett.* **2010**, *1*, 912–917. [CrossRef]
3. Couture, R.; Rose, J.; Kumar, N.; Mitchell, K.; Wallschläger, D.; Cappellen, P. Sorption of Arsenite, Arsenate, and Thioarsenates to Iron Oxides and Iron Sulfides: A Kinetic and Spectroscopic Investigation. *Environ. Sci. Technol.* **2013**, *47*, 5652–5659. [CrossRef] [PubMed]
4. Du, J.; Bao, J.; Lu, C.; Werner, D. Reductive sequestration of chromate by hierarchical FeS@Fe 0 particles. *Water Res.* **2016**, *102*, 73–81. [CrossRef] [PubMed]

5. Guan, X.; Sun, Y.; Qin, H.; Li, J.; Lo, I.; Di, H. The limitations of applying zero-valent iron technology in contaminants sequestration and the corresponding countermeasures: The development in zero-valent iron technology in the last two decades (1994–2014). *Water Res.* **2015**, *75*, 224–248. [CrossRef] [PubMed]

6. Han, Y.; Yan, W. Reductive Dechlorination of Trichloroethene by Zero-valent Iron Nanoparticles: Reactivity Enhancement through Sulfidation Treatment. *Environ. Sci. Technol.* **2016**, *50*, 12992–13001. [CrossRef] [PubMed]

7. Boente, C.; Sierra, C.; Martinez-Blanco, D.; Menendez-Aguado, J.M.; Gallego, J.R. Nanoscale zero-valent iron-assisted soil washing for the removal of potentially toxic elements. *J. Hazard. Mater.* **2018**, *350*, 55–65. [CrossRef] [PubMed]

8. Liu, X.J.; Lai, D.G.; Wang, Y. Performance of Pb(II) removal by an activated carbon supported nanoscale zero-valent iron composite at ultralow iron content. *J. Hazard. Mater.* **2019**, *361*, 37–48. [CrossRef] [PubMed]

9. Mojiri, A.; Kazeroon, R.A.; Gholami, A. Cross-Linked Magnetic Chitosan/Activated Biochar for Removal of Emerging Micropollutants from Water: Optimization by the Artificial Neural Network. *Water* **2019**, *11*, 551. [CrossRef]

10. Jia, Z.; Shu, Y.; Huang, R.; Liu, J.; Liu, L. Enhanced reactivity of nZVI embedded into supermacroporous cryogels for highly efficient Cr(VI) and total Cr removal from aqueous solution. *Chemosphere* **2018**, *199*, 232–242. [CrossRef] [PubMed]

11. Li, J.; Chen, C.; Zhu, K.; Wang, X.J. Nanoscale zero-valent iron particles modified on reduced graphene oxides using a plasma technique for Cd(II) removal. *J. Taiwan Inst. Chem. Eng.* **2016**, *59*, 389–394. [CrossRef]

12. Lai, B.; Chen, Z.; Zhou, Y.; Yang, P.; Wang, J.; Chen, Z. Removal of high concentration p -nitrophenol in aqueous solution by zero valent iron with ultrasonic irradiation (US–ZVI). *J. Hazard. Mater.* **2013**, *250*, 220–228. [CrossRef] [PubMed]

13. Chen, Z.X.; Jin, X.Y.; Chen, Z.; Megharaj, M.; Naidu, R. Removal of methyl orange from aqueous solution using bentonite-supported nanoscale zero-valent iron. *J. Colloid Interface Sci.* **2011**, *363*, 601–607. [CrossRef] [PubMed]

14. Sheng, G.; Shao, X.; Li, Y.; Li, J.; Dong, H.; Cheng, W.; Gao, X.; Huang, Y. Enhanced Removal of Uranium(VI) by Nanoscale Zerovalent Iron Supported on Na-Bentonite and an Investigation of Mechanism. *J. Phys. Chem. A* **2014**, *118*, 2952–2958. [CrossRef] [PubMed]

15. Wang, J.; Liu, G.; Li, T.; Zhou, C.; Qi, C. Zero-Valent Iron Nanoparticles (NZVI) Supported by Kaolinite for CuII and NiII Ion Removal by Adsorption: Kinetics, Thermodynamics, and Mechanism. *Aust. J. Chem.* **2015**, *68*, 1305–1315. [CrossRef]

16. Zhu, S.; Huang, X.; Wang, D.; Wang, L.; Ma, F. Enhanced hexavalent chromium removal performance and stabilization by magnetic iron nanoparticles assisted biochar in aqueous solution: Mechanisms and application potential. *Chemosphere* **2018**, *207*, 50–59. [CrossRef] [PubMed]

17. Fu, F.; Ma, J.; Xie, L.; Tang, B.; Han, W.; Lin, S. Chromium removal using resin supported nanoscale zero-valent iron. *J. Environ. Manag.* **2013**, *128*, 822–827. [CrossRef]

18. Huang, T.; Liu, L.; Zhou, L.; Zhang, S. Electrokinetic removal of chromium from chromite ore-processing residue using graphite particle-supported nanoscale zero-valent iron as the three-dimensional electrode. *Chem. Eng. J.* **2018**, *350*, 1022–1034. [CrossRef]

19. Vilardi, G.; Mpouras, T.; Dermatas, D.; Verdone, N.; Polydera, A.; Di Palma, L. Nanomaterials application for heavy metals recovery from polluted water: The combination of nano zero-valent iron and carbon nanotubes. Competitive adsorption non-linear modeling. *Chemosphere* **2018**, *201*, 716–729. [CrossRef]

20. Kim, H.; Hong, H.J.; Jung, J.; Kim, S.H.; Yang, J.W. Degradation of trichloroethylene (TCE) by nanoscale zero-valent iron (nZVI) immobilized in alginate bead. *J. Hazard. Mater.* **2010**, *176*, 1038–1043. [CrossRef]

21. Ma, Q.; Zhang, H.; Deng, X.; Cui, Y.; Cheng, X.; Li, X. Electrochemical fabrication of NZVI/TiO2 nano-tube arrays photoelectrode and its enhanced visible light photocatalytic performance and mechanism for degradation of 4-chlorophenol. *Sep. Purif. Technol.* **2017**, *182*, 144–150. [CrossRef]

22. Chen, S.; Bedia, J.; Li, H.; Ren, L.Y.; Naluswata, F.; Belver, C. Nanoscale zero-valent iron@mesoporous hydrated silica core-shell particles with enhanced dispersibility, transportability and degradation of chlorinated aliphatic hydrocarbons. *Chem. Eng. J.* **2018**, *343*, 619–628. [CrossRef]

23. Gu, M.; Farooq, U.; Lu, S.; Zhang, X.; Qiu, Z.; Sui, Q. Degradation of trichloroethylene in aqueous solution by rGO supported nZVI catalyst under several oxic environments. *J. Hazard. Mater.* **2018**, *349*, 35–44. [CrossRef] [PubMed]

24. Yang, L.; Gao, J.; Liu, Y.; Zhang, Z.; Zou, M.; Liao, Q.; Shang, J. Removal of Methyl Orange from Water Using Sulfur-Modified nZVI Supported on Biochar Composite. *Water Air Soil Pollut.* **2018**, *229*, 355. [CrossRef]
25. Zhou, W.H.; Liu, F.; Yi, S.; Chen, Y.Z.; Geng, X.; Zheng, C. Simultaneous stabilization of Pb and improvement of soil strength using nZVI. *Sci. Total Environ.* **2019**, *651*, 877–884. [CrossRef] [PubMed]
26. Barzegar, G.; Jorfi, S.; Zarezade, V.; Khatebasreh, M.; Mehdipour, F.; Ghanbari, F. 4-Chlorophenol degradation using ultrasound/peroxymonosulfate/nanoscale zero valent iron: Reusability, identification of degradation intermediates and potential application for real wastewater. *Chemosphere* **2018**, *201*, 370–379. [CrossRef]
27. Hu, S.; Hu, J.; Liu, B.; Wang, D.; Wu, L.; Xiao, K.; Liang, S.; Hou, H.; Yang, J. In situ generation of zero valent iron for enhanced hydroxyl radical oxidation in an electrooxidation system for sewage sludge dewatering. *Water Res.* **2018**, *145*, 162–171. [CrossRef]
28. Karpov, M.; Seiwert, B.; Mordehay, V.; Reemtsma, T.; Polubesova, T.; Chefetz, B. Transformation of oxytetracycline by redox-active Fe(III)- and Mn(IV)-containing minerals: Processes and mechanisms. *Water Res.* **2018**, *145*, 136–145. [CrossRef]
29. Ravikumar, K.V.G.; Sudakaran, S.; Nancharaiah, S.V.; Chandrasekaran, N.J. Biogenic nano zero valent iron (Bio-nZVI) anaerobic granules for textile dye removal. *Environ. Chem. Eng.* **2018**, *6*, 1683–1689.
30. Li, X.G.; Zhao, Y.; Xi, B.D.; Meng, X.G.; Gong, B.; Li, R.; Peng, X.; Liu, H.L. Decolorization of Methyl Orange by a new clay-supported nanoscale zero-valent iron: Synergetic effect, efficiency optimization and mechanism. *J. Environ. Sci. China* **2017**, *52*, 8–17. [CrossRef]
31. He, J.; Li, Y.; Cai, X.; Chen, K.; Zheng, H.; Wang, C.; Zhang, K.; Lin, D.; Kong, L.; Liu, J. Study on the removal of organic micropollutants from aqueous and ethanol solutions by HAP membranes with tunable hydrophilicity and hydrophobicity. *Chemosphere.* **2017**, *174*, 380–389. [CrossRef] [PubMed]
32. Wang, T.; Su, J.; Jin, X.; Chen, Z.; Naidu, R. Functional clay supported bimetallic nZVI/Pd nanoparticles used for removal of methyl orange from aqueous solution. *J. Hazard. Mater.* **2013**, *262*, 819–825. [CrossRef] [PubMed]
33. Li, D.; Sun, T.; Wang, L.; Wang, N. Enhanced electro-catalytic generation of hydrogen peroxide and hydroxyl radical for degradation of phenol wastewater using MnO_2/Nano-G|Foam-Ni/Pd composite cathode. *Electrochim. Acta* **2018**, *282*, 416–426. [CrossRef]
34. Mameda, N.; Park, H.; Choo, K.H. Electrochemical filtration process for simultaneous removal of refractory organic and particulate contaminants from wastewater effluents. *Water Res.* **2018**, *144*, 699–708. [CrossRef] [PubMed]
35. Mao, X.; Tian, W.; Ren, Y.; Chen, D.; Curtis, S.E.; Buss, M.T.; Rutledge, G.C.; Hatton, T.A. Energetically efficient electrochemically tunable affinity separation using multicomponent polymeric nanostructures for water treatment. *Energy Environ. Sci.* **2018**, *11*, 2954–2963. [CrossRef]
36. Shen, J.; Ou, C.; Zhou, Z.; Chen, J.; Fang, K.; Sun, X.; Li, J.; Zhou, L.; Wang, L. Pretreatment of 2,4-dinitroanisole (DNAN) producing wastewater using a combined zero-valent iron (ZVI) reduction and Fenton oxidation process. *J. Hazard. Mater.* **2013**, *260*, 993–1000. [CrossRef] [PubMed]
37. Shan, D.; Deng, S.; Jiang, C.; Chen, Y.; Wang, B.; Wang, Y.; Huang, J.; Yu, G.; Wiesner, M.R. Hydrophilic and strengthened 3D reduced graphene oxide/nano-Fe_3O_4 hybrid hydrogel for enhanced adsorption and catalytic oxidation of typical pharmaceuticals. *Environ. Sci. Nano* **2018**, *5*, 1650–1660. [CrossRef]

Article

Comparative Study of Four TiO$_2$-Based Photocatalysts to Degrade 2,4-D in a Semi-Passive System

Gisoo Heydari [1,*]**, Jordan Hollman** [1]**, Gopal Achari** [1] **and Cooper H. Langford** [2]

[1] Department of Civil Engineering, University of Calgary, 2500 University Dr. NW, Calgary, AB T2N 1N4, Canada; jordan.hollman@ucalgary.ca (J.H.); gachari@ucalgary.ca (G.A.)

[2] Department of Chemistry, University of Calgary, 2500 University Dr. NW, Calgary, AB T2N 1N4, Canada; chlangfo@ucalgary.ca

* Correspondence: gisoo.heydari@ucalgary.ca

Received: 14 February 2019; Accepted: 21 March 2019; Published: 26 March 2019

Abstract: In this study, the relative efficiency of four forms of supported titanium dioxide (TiO$_2$) as a photocatalyst to degrade 2,4-dichlorophenoxyacetic acid (2,4-D) in Killex®, a commercially available herbicide was studied. Coated glass spheres, anodized plate, anodized mesh, and electro-photocatalysis using the anodized mesh were evaluated under an ultraviolet – light-emitting diode (UV-LED) light source at $\lambda = 365$ nm in a semi-passive mode. Energy consumption of the system was used to compare the efficiency of the photocatalysts. The results showed both photospheres and mesh consumed approximately 80 J/cm^3 energy followed by electro-photocatalysis (112.2 J/cm^3), and the anodized plate (114.5 J/cm^3). Although electro-photocatalysis showed the fastest degradation rate (K = 5.04 mg L^{-1} h^{-1}), its energy consumption was at the same level as the anodized plate with a lower degradation rate constant of 3.07 mg L^{-1} h^{-1}. The results demonstrated that three-dimensional nanotubes of TiO$_2$ surrounding the mesh provide superior degradation compared to one-dimensional arrays on the planar surface of the anodized plate. With limited broad-scale comparative studies between varieties of different TiO$_2$ supports, this study provides a comparative analysis of relative degradation efficiencies between the four photocatalytic configurations.

Keywords: photocatalysis; semi-passive; anodization; buoyant catalyst; 2,4-D; LED; mesh

---

1. Introduction

As a clean technology, photocatalysis holds a lot of promise. Research interest on photocatalysis has significantly increased since the initial publication on titanium dioxide (TiO$_2$) photocatalysis in 1971 [1]. As an advanced oxidation process, it utilizes a semiconductor material as a reusable catalyst that is capable of mineralizing organic contaminants with light energy as the main input. This makes photocatalysis an appealing treatment option for both industrial and municipal wastewaters [2–7].

The most versatile semiconductor that has been used as a photocatalyst is titanium dioxide (TiO$_2$), which owes its popularity to its low cost, non-toxicity, and photo-stability as well as its unique non-selectivity characteristic for oxidation reactions. It is also chemically and mechanically robust, that makes it an ideal photocatalyst candidate for various reaction media [8–11]. With a high oxidation potential (approximately +3.2 V), high energy ultraviolet (UV) radiation ($\lambda \leq 387$ nm) is required to excite electrons in its valence band and move them to the conduction band to initiate the photocatalytic reaction [4,9,12]. Although other semiconductors such as zinc oxide (ZnO), cadmium sulfide (CdS) [13], tungsten oxides (WOx $\leq$ 3) [14,15] and other tungstate species [16] as well as their combinations [2,4,17] have been tested for photocatalysis, TiO$_2$ is still a broadly studied photocatalyst and is utilized as a model photocatalyst in this study.

TiO$_2$ can be used in various forms such as powder, being immobilized (on a surface such as a sphere or a plate), or one of many possible variations of nanoparticle materials. Each method of application presents unique advantages and challenges. While powdered TiO_2 has been extensively researched, it has a significant disadvantage in environmental applications. As a photoactive material, the powdered TiO_2 must be separated from water prior to its release that necessitates an additional treatment step [3,18,19]. Although immobilized TiO_2 avoids the complications of a secondary separation step, its available surface for reactions is reduced and consequently has lower reactivity, leading to lower degradation rates in some cases [20–23].

Supported nanostructures are a viable option for the application of TiO_2 in water treatment. They can provide higher reaction rates than simple immobilized TiO_2 plates while having the advantage of being supported, making an extra separation step unnecessary. Many studies on TiO_2 nanotubes boast high electron mobility, lower recombination rates, high surface area and high mechanical strength [24–29]. When nanotubes are produced on spaced materials such as a mesh, nanotubes can grow in all directions resulting in a three-dimensional structure. Several studies have been published demonstrating degradation of contaminants using these three-dimensional nanotube structures [26,30,31]. Research [26] has demonstrated that photocatalyst nanotubes with a 3D geometry were more efficient in absorbing light, minimizing photon loss in the liquid, and present a much higher photocatalytic activity per unit surface area compared to a plated one-dimensional array. The results showed that the photocurrent response in the mesh, as an indicator of the photocatalytic activity, was higher than the plate. This indicated a lower recombination of photo-generated electrons and holes, higher photoelectron transfer efficiency and higher light absorbance in the mesh. The increased light absorbance efficiency can be attributed to the different directionality of nanotubes with 3D geometry being more effective in capturing indirect light such as reflected or refracted photons. Beyond the improved light absorbance, better degradation rates in 3D structures have also been attributed to interstitial fissures between the nanotubes, which provides more access to the catalyst surface by the contaminant [26].

Amongst various fabrications and application methods, positioning a photocatalyst near the surface of water, and on a floating support have been investigated by several researchers as it allows higher oxygenation and illumination [17,32–36]. Although the initial floating supported photocatalysts were shown to be effective to clean oil slicks on water [17], the ingestion risk by fishes and animals in water and uncertain toxicity of the degradation intermediates limited their development. This led to various studies to improve their size scale and evaluate their efficacy to degrade various contaminants. Alternatively, TiO_2 plates are a commonly considered photocatalyst. With a simple setup, they have been shown to degrade a variety of organic contaminants [8,37]. However, due to the tendency of charge-hole recombination in the semiconductor photocatalyst, the quantum yield is generally low in heterogeneous photocatalysis [19,20,38].

One of the methods to overcome the low quantum yield of heterogeneous photocatalysis on supported substrates is electro-photocatalysis. This refers to an anodic polarization being applied in an electrochemical cell, causing it to act as a photoanode. The applied voltage removes excited electrons from the surface of the photocatalyst, inhibiting the recombination of the electron-hole pairs [39,40]. Researches have demonstrated the enhanced rate of photocatalytic degradation when it is combined with an electrical bias. Research on an Azo dye showed that the degradation rate approximately doubled when an electrical bias of 1.5 V was applied on the nanotube arrays of an anodized Ti mesh electrode [41]. TiO_2 coated electrode by the sol-gel method resulted in 65% enhancement in degradation efficiency of methyl tertiary butyl ether using 0.25 bias [42], and the electro-photocatalytic degradation rate of methyl orange was 1.7 times higher on the modified titanium nanotube electrode in comparison to the photocatalytic degradation [43].

There is extensive literature investigating various supports for photocatalysis. Most studies provide a comparison between different nanotube structures or various coating methods, without an experimental peer evaluation between the efficiency of different supports. Considering the advantages of buoyant photocatalysts and nanostructures, the intention of this research was to conduct an exploratory study to compare the efficacy between the conventional floating TiO_2 coated on the glass spheres and nanostructured engineered supports, to be used as floating photocatalysts for water decontamination under ambient conditions.

In this research, an experimental evaluation of four types of supported TiO_2 photocatalysts were conducted, allowing for a broad comparison between the different supports in a semi-passive mode. Floating TiO_2 spheres, an anodized TiO_2 plate with one-dimensional nanotube arrays, anodized TiO_2 mesh with a three-dimensional nanotube structure, and electro-photocatalysis utilizing the anodized TiO_2 mesh were investigated.

The herbicide Killex® was used as a model organic contaminant. It is a widely applied herbicide that is used in lawns and agricultural lands. It contains 2,4-dichlorophenoxyacetic acid (2,4-D), methylchlorophenoxy propionic acid (MCPP or Mecoprop-P) and 3,6-dichloro-2-methoxybenzoic acid (Dicamba). 2,4-D is a contaminant of priority due to its low biodegradability and runoff potential and high mobility [44,45]. Photocatalytic degradation of 2,4-D results in the production of 4-chloro pyrocatechol, 2-chlorophenol, 4-chlorophenol, 2,4-dichlorophenol [46–48]. During photocatalytic degradation of Killex®, degradation of 2,4-D is slower compared to its pure aqueous solution due to the competition between its three components [48,49].

The irradiation source was an ultraviolet – light-emitting diode (UV-LED) panel. UV-LEDs were selected for their energy efficiency and longer lifespan compared to low-pressure UV lamps (i.e., approximately four times) [6,49,50]. Additionally, LEDs contain no mercury, making them an environmentally friendly alternative to standard fluorescent UV bulbs.

The experiments were conducted in a semi-passive mode, where no mixing was involved during the irradiation. The objective was to investigate what can be expected of a semi-passive photocatalytic system if it was to be used to treat ponds containing contaminated water under an ambient environment. Thereby, the reaction rates can be representative of the rate of removal of organic contaminants in natural contaminated watersheds and industrial wastewater ponds that are exposed to sunlight.

2. Materials and Methods

2.1. Materials

The titanium mesh with 99.9% purity was supplied from Alfa Aesar. The titanium plate with 99.7% purity was obtained from Sigma-Aldrich, St. Louis, MO, USA. The 6061-grade aluminum was used as a cathode for anodization and was procured from New West Metals Inc., Winnipeg, MB, Canada. The hollow glass microspheres coated with anatase TiO_2 (HGMT) were obtained from Cospheric Innovations in Microtechnology. They are referred to as photospheres in this paper and had a median diameter of 45 μm and density of 0.22 g/cm^3.

A 99% pure-ethylene glycol was supplied by VWR and 99.9% pure acetone was obtained from EMD. A 98% pure ammonium fluoride, 99.8% ethyl acetate and 99.8% methanol were obtained from Sigma-Aldrich. The phosphoric acid was supplied by J. T. Baker. Commercially available Killex® containing 95 g/L of 2,4-dichlorophenoxyacetic acid (2,4-D), 52.5 g/L of Mecoprop-P, and 9 g/L of Dicamba was purchased from a local retailer.

An in-house UV-LED light source (λ = 365 nm) composed of 16 lamps (NSCU330B, Nichia Corporation, Anan, Japan) in a four by four array on a 10 cm by 10 cm circuit board was used as a light source. It was equipped with an air fan to cool down the generated heat from the lamps. A dual output power supply (Model TW 5005D) was used to drive the module with direct current, ranging from 10 mA to 500 mA [51]. A PANalytical Aeris X-ray diffractometer was used for X-ray diffraction (XRD) characterization of the photocatalysts. The HighScore Plus XRD analysis software was utilized for Qualitative XRD analysis and Rietveld Refinement. An SEM (JEOL JSM-IT300LV, Tokyo, Japan) by InTouchScope and a TEM (Tecnai F20, by Thermo Fischer Scientific) were used to characterize the morphologies of the anodized titanium mesh. A Hummer II sputter coater from Anatech, USA was used to coat the cathode (anodized titanium mesh) with platinum.

2.2. Methods

2.2.1. Experimental Design and Setup

All experiments were conducted at room temperature, pressure, and under similar conditions. In order to be able to compare the photocatalyst supports (as a variable), all experimental parameters including light source, irradiation light intensity, lack of mixing, distance from the light source and the initial concentration of 2,4-D in the solution were identical in all experiments. The type of the photocatalyst support was different in each experiment, therefore the performance of the studied photocatalysts (i.e., TiO_2 photospheres, anodized titanium mesh, plate and electro-photocatalysis using an anodized titanium mesh) were studied as a variable.

The output light intensity of the UV-LED module was linear in the range of 2.2 mW/cm^2 to 17.3 mW/cm^2. The light intensity was adjusted at 15 mW/cm^2 [51]. The distance of the photocatalyst surface from the irradiation source was 5.5 cm in all experiments.

Wide mounted glass jars with a diameter of 4 cm were used as reaction vessels. In all experiments, 0.52 mL of Killex$^®$ was dissolved in 1 L of milli-Q water resulting in a solution containing 49.4 ppm 2,4-D, 27.3 ppm Mecoprop-P and 4.68 ppm Dicamba. In each experiment, a control sample without a photocatalyst was also monitored to evaluate the possible photolysis in the solutions. The length of time between sampling intervals created a requirement to correct the mass of the solution for evaporation. This was done by monitoring the mass of the solution during the experiments. The milli-Q water was added before each sampling to adjust the mass of the solution to its initial mass before each irradiation cycle.

In the experiments using an anodized titanium plate, the plate was placed at the bottom of the reaction jar. Dimensions of the plate were measured and used to calculate the surface area of the photocatalyst for energy calculations. When the titanium mesh was used, it was secured close to the surface of the reaction medium using a stainless steel wire mesh. The surface area of the mesh was calculated based on its dimensions and the percentage of the open area (64%) was used to calculate the effective surface area for energy calculations.

In the electro-photocatalysis experiments, a platinum coated anodized mesh was used as a cathode, and the anodized titanium mesh was used as a photoanode. Anode and cathode were connected to a battery (1.5 V) as a source of direct electric current. Current and voltage were measured with a voltmeter and ammeter during the experiment.

For the TiO_2 photospheres experiments, the photospheres were mixed with water and those that were buoyant, were collected and dried overnight. The photospheres were then mixed with spiked water using a magnetic stirrer for 30 min before irradiation to allow adsorption to reach equilibrium. In order to identify an optimum photocatalyst loading for the purpose of comparison with the other studied photocatalysts, three different catalyst loadings were tested, 7.2 mg/cm^2 and 11.9 mg/cm^2 and 16.7 mg/cm^2, equal to 1.8 g/L, 3.0 g/L, and 4.2 g/L, respectively. Since the photocatalyst was floating on the surface, the surface area of the reaction jar was considered as the irradiated surface by the light source and was used in energy calculations.

2.2.2. Anodization of Titanium Mesh

Supported TiO_2 nanotubes were prepared by anodization of a titanium mesh followed by a temperature controlled annealing to acquire anatase phase, which is the most desired crystalline morphology of TiO_2 for photocatalytic applications [52]. The titanium mesh was first degreased by sonication in an acetone/methanol solution for 30 min. It was then rinsed with water and dried at room temperature. The degreased mesh was anodized in an aqueous solution of ethylene glycol, containing 2% water and 0.5% ammonium fluoride for 24 h. The anodized mesh was then rinsed with water and dried at room temperature followed by annealing in the oven for 3 h at 450 °C to form an anatase crystalline structure [24,26,45–47]. The mesh was cut into uniform pieces with a surface area of

2.9 $\pm$ 0.3 cm^2, and each of them were attached to a reaction jar via a stainless steel wire and irradiated under the UV-LED lamp.

2.2.3. Sample Analysis

Samples were filtered using 0.45 µm PTFE 25 mm syringe filters before analysis, and were analyzed using HPLC with a UV-visible detector, fitted with a Restek Pinnacle DB C18 column. Absorbance was recorded at a wavelength of 230 nm. Ingredients of Killex$^®$, have all the peak wavelength at the range of 230 nm to 280 nm (i.e., 2,4-D: λ_{max} = 280 nm, Mecoprop: λ_{max} = 230 nm and Dicamba: λ_{max} = 275 nm) [6,53,54]. The UV detection wavelength of 230 nm is selected so as to be able to detect all three compounds in the Killex$^®$ solution [55]. However, degradation of 2,4-D as a model contaminant is studied in this paper for the purpose of comparison of the photocatalysts. The eluent used was water:methanol (25:75) with ten mM phosphoric acid and a flow rate of 1 L/min.

3. Results and Discussion

3.1. Photocatalyst Characterization

Titanium dioxide exists in three different crystalline structures of rutile, anatase and brookite [56]. The anatase and rutile show different structural characteristics in terms of Ti–Ti and Ti–O distances. The Ti–Ti distance is larger in anatase than rutile but the Ti–O distance is shorter. These structural differences lead to different mass densities and different electronic structures that causes the difference in the mobility of the charge carriers under light excitation [57]. Amongst all crystal structures of TiO$_2$, the anatase shows the highest photocatalytic activity [56]. The X-ray diffractometry (XRD) of the photocatalysts indicated that the primary crystal structure of all the photocatalysts was anatase. (The XRD graphs of all the photocatalyst samples are presented in the supplementary material).

The XRD quantitative results that show the weight percentage of titanium, TiO$_2$, and its crystalline morphologies are presented in Table 1.

Table 1. XRD quantitative results of titanium dioxide (TiO$_2$) photocatalysts.

Photocatalyst	Crystal Phase (weight %)		
	Anatase	Rutile	Titanium
Mesh	91.8	3.7	2.8
Plate	81.9	0.1	17.3
Photospheres	72.7	-	-

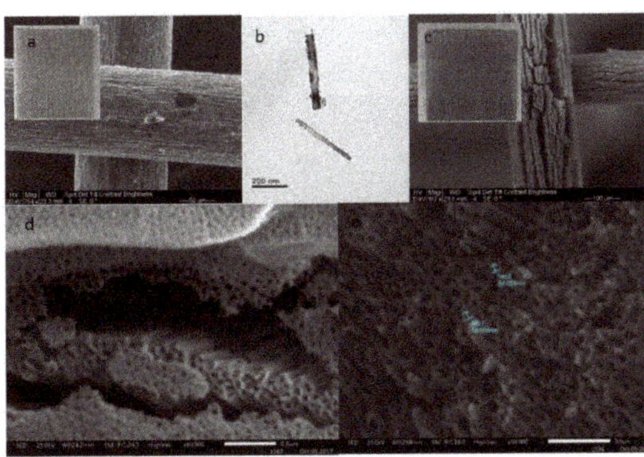

Figure 1. Morphology of the anodized mesh: (**a**) Digital and SEM images before anodization; (**b**) TEM image of a nanotube detached from the surface; (**c**) digital and SEM images after anodization; (**d,e**) SEM image of titanium nanotube arrays on the surface of mesh.

Further analysis was performed on the mesh photocatalyst for characterizing the surface morphology, length, and thickness of the nanostructures by a scanning electron microscope (SEM) and a transmission electron microscope (TEM). Images are presented in Figure 1. As it can be seen on the images, the length of the nanotubes was about 300 nm with the opening diameter of 54 nm.

3.2. TiO$_2$ Photospheres

A comparative study of the photosphere loadings was conducted. The results are presented in Figure 2. Samples were irradiated for nine h and 22 min with a total energy of 127.1 J/cm^3. The 2,4-D has no absorbance at the peak wavelength of LED (λ = 365 nm). Therefore, no direct photolysis took place during its degradation [6]. The degradation showed a notable improvement when increased beyond a photosphere loading of 7.2 mg/cm^2, while results of 11.9 mg/cm^2 and 16.7 mg/cm^2 were similar. The degradation by photospheres in this study followed a zero-order reaction kinetics, with the kinetic rate constant (K value) of 4.12 mg L^{-1} h^{-1} and the half-life time of 6.05 h at a loading of 7.2 mg/cm^2. K values were 4.55 mg L^{-1} h^{-1} and 4.36 mg L^{-1} h^{-1}, and the half-life times were 5.48 h and 5.72 h at the two higher loadings of 11.9 mg/cm^2 and 16.7 mg/cm^2 respectively. No degradation took place in the control sample. Zero-order reaction kinetics was also reported in other studies, where TiO$_2$ floating beads were used for the degradation of dyes [36]. The degradation efficiency was 80%, 86% and 89% at the loadings of 7.2 mg/cm^2, 11.9 mg/cm^2 and 16.7 mg/cm^2 respectively. The results are summarized in Table 2.

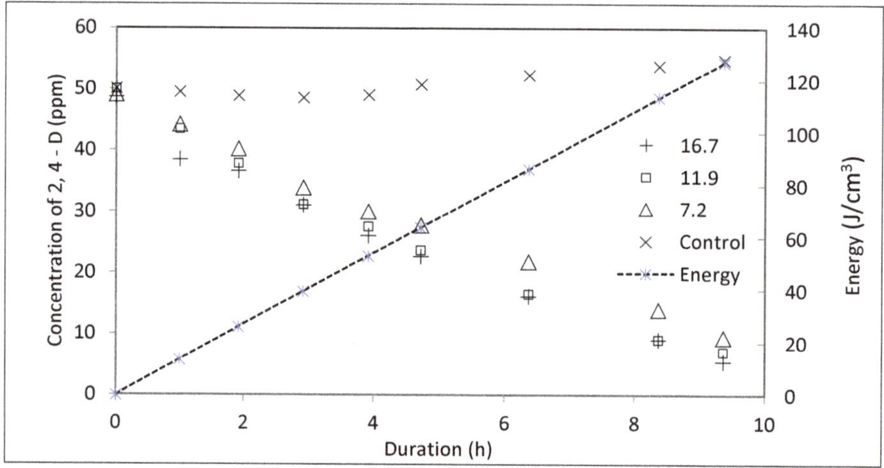

Figure 2. Degradation of 2,4-dichlorophenoxyacetic acid (2,4-D) in Killex® solution at various loadings of TiO$_2$ photospheres (mg/cm^2).

Table 2. Summary of the experimental results using photospheres.

Photocatalyst Loading (mg/cm^2)	K (mg L^{-1} h^{-1})	$t_{\frac{1}{2}}$ (h)	Degradation (%)
7.2	4.12	6.05	80
11.9	4.55	5.48	86
16.7	4.36	5.72	89

The results indicated that increasing the loading of photospheres in the solution increased the degradation rate up to an optimum value beyond which the degradation rate decreased as more photocatalyst was added. This phenomenon has been demonstrated by various researchers [58]. Since heterogeneous photocatalysis occurs inside the active sites on the surface of TiO$_2$, the rate of degradation increases when the loading of the photocatalyst increases. This is due to the higher surface area, hence the availability of the active reaction sites. However, an optimum TiO_2 dosage exists after which degradation rates will decrease with higher catalyst loadings [59]. This phenomenon has been attributed to photon limitation or light scattering on the surface of the photocatalyst particles, decreased light penetration through active sites, and uneven competition between the nanoparticles for light adsorption at higher loadings [60]. Moreover, the high catalyst loading may deactivate the originally activated TiO_2 through collision with ground state catalysts. The following reaction demonstrates the collision mechanism between TiO_2 species. TiO_2^* has the activated species adsorbed on its surface, TiO_2 is the ground state and $TiO_2^\#$ is the deactivated state of TiO_2 [61,62]. When the light activated TiO_2^* collides with the ground state TiO_2, TiO_2^* gets deactivated ($TiO_2^\#$), leaving the ground state behind. A high loading of the photocatalyst, increases the chance of collision between the species, resulting in the higher frequency in the deactivation process.

$$TiO_2^* + TiO_2 \rightarrow TiO_2^\# + TiO_2, \tag{1}$$

A screening effect of UV irradiation is also reported to be responsible for lowering the rate of photocatalysis at the loading values beyond the threshold limit [59,63]. In addition, due to the uncontrolled catalyst agglomeration on the sidewalls of the reaction jar, increasing the catalyst loading does not enhance the available catalyst on the surface of the reaction media. Since most of the studies utilize photospheres in a reactor with mixing, this phenomenon is fairly reported in the literature [9,64].

The experiments resulted in similar degradation rates in two photocatalyst loadings. Therefore, it is possible that the optimum level of loading was reached. This allows for a comparison to other suspended photocatalytic methods as the measured level of treatment represents the highest level of attainable degradation by photospheres.

3.3. Anodized Mesh

An experimental study utilizing the anodized TiO_2 mesh was performed to assess its photocatalytic activity in a semi-passive mode, and as well to compare it with the anodized plate and other methods studied here. The degradation of 2,4-D with time is presented in Figure 3. The zero-order kinetic rate constant was 3.45 mg L^{-1} h^{-1} with the half-life time ($t_{1/2}$) of 7.09 h. Energy consumption to achieve 99.8% degradation was on average 160 J/cm^3.

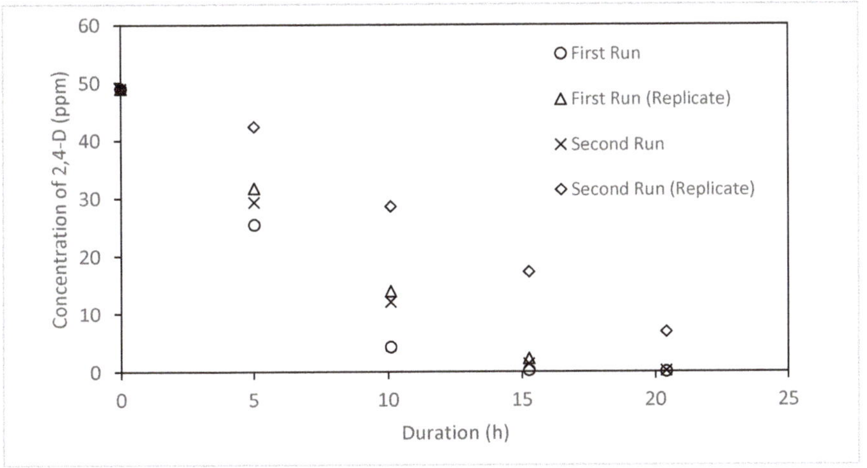

Figure 3. Degradation of 2,4-D using anodized mesh.

Experiments were performed in duplicate to account for considerable variances that can occur when growing TiO_2 nanotubes on the non-uniform surface of the mesh. The degradation efficiency varied 2.2% between the replicate samples. Although the percentage was small, the difference between samples can be attributed to the variation on nanotube structure due to the variations in the electrical field alongside the wires of the mesh during the anodization process. This emphasizes the importance of the production of a uniform mesh.

Each TiO_2 mesh was tested a second time to assess its durability in repeated applications. One of the samples showed a similar efficiency with degradation efficiency varied between 99.65 to 100%, but the degradation efficiency dropped 13% in the second replicate. The decreased efficiency has been reported for repeated runs on TiO_2 nanotube meshes by Liao et al. [26], who found that after five runs the degradation efficiency decreased by 8%. The mechanism for loss of degradation efficiency is due to the imperfect mechanical stability of the nanotube layers as TiO_2 nanotubes tend to flake off from the substrate during the experiment [65]. Previous research [66] showed that the mechanical stability of TiO_2 nanotube arrays could be improved by surface modification methods such as incorporating carbon into microstructures of nanotube layers, which increases the hardness and mechanical strength of the layers. In another study, it was demonstrated that the presence of fluoride-rich layer improves the adhesion of the nanotube layer [65].

Variables affecting anodization are electrolytes types, voltage, and duration of the anodization, annealing temperature and duration, as well as the distance between the electrodes. They cause a difference in the crystal phase, length and width of the produced TiO_2 nanotubes which is the reason behind the difference in their mechanical stability on the surface of the titanium mesh [25,26,28,67–70]. When the nanotubes are not mechanically stable, they can fall off from the surface in repetitive usage, resulting in a reduced photocatalytic activity.

3.4. Anodized Plate

The degradation efficiency was 57% after 10 h of irradiation, consuming 111.9 J/cm^3 of energy. Degradation followed a zero-order degradation rate constant of 3.07 mg L^{-1} h^{-1} with the half-life time of 7.97 h. Since the photocatalyst plate was placed at the bottom of the reaction jar, an additional experiment was performed to investigate the effect of the depth of water column on the degradation efficiency. The plate was covered with contaminated water, which resulted in various water depths ranging from 1 cm to 4 cm on top of the photocatalyst plate. The concentration, distance from the irradiation source and the light intensity were identical in all samples. All samples irradiated for 10 h. The results are presented in Figure 4.

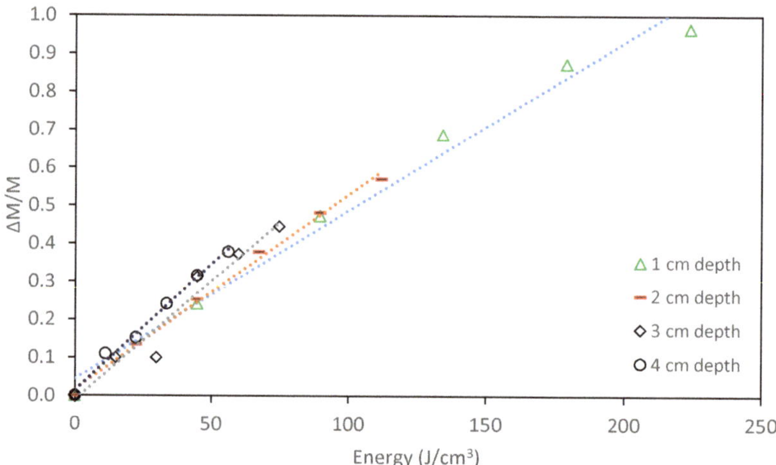

Figure 4. Degradation efficiency versus energy at various depths using the anodized plate.

It was observed that at a depth of 4 cm, energy consumption was 55.9 J/cm^3 to achieve 38% degradation efficiency (defined as $\Delta M/M_0 = (M_0 - M_t)/M_0$), which was 27% lower compared to a 1 cm depth (76.3 J/cm^3). The results showed that the increase in depth, did not decrease the energy efficiency of the system. Although the Beer Lambert equation (A = ε b c) states light absorbance in a solution (A) is directly proportional to its path length (b), molar concentration (c), and molar absorptivity (ε), the results showed that the effect of path length at the studied depth ranging from 1 cm to 4 cm was negligible.

This could be associated with the mass of the contaminant which is a significant parameter in the photocatalytic degradation reactions due to its effect on the adsorption of the reactant on the surface of the photocatalyst and mass transfer in the solution during the photocatalytic degradation reaction [9,71,72]. Since the mass was higher when the water column was higher, the effect of mass supersedes the effect of light absorbance between the studied depths of 1 to 4 cm. It was then concluded that when the anodized plate is placed at the bottom of the reaction jar, the effect of the depth of water column is not as important as the mass of the contaminant; hence it can be neglected. Therefore, the results are comparable with the floating photospheres and mesh, where the photocatalyst is brought near the surface.

3.5. Electro-Photo Catalysis Using Anodized Titanium Mesh

A study for comparative analysis of electro-photocatalysis was conducted on the anodized TiO$_2$ mesh. Voltage, as the driving force in preventing electron-hole pair recombination [39], was kept constant at 1.5 $\pm$ 0.1 V during the 6 h irradiation. The photocurrent increased from 6 mA before irradiation to about 40 mA after UV irradiation. Voltage was variable at the first 20 min which was corrected by manually adjusting to the initial value of 1.5 V. The current was approximately 45 mA during the 6 h irradiation period. Results of the experiment are presented in Figure 5.

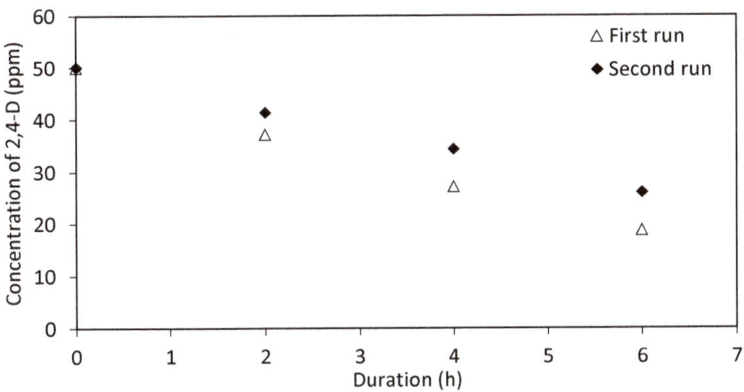

Figure 5. Concentration of 2,4-D with time during electro-photocatalysis.

Kinetics followed a zero-order model with a rate constant of 5.04 mg L^{-1} h^{-1} with the half-life time of 4.86 h during the first run. The enhanced degradation efficiency is due to the presence of the polarized anode, working as a photoanode in the photoelectrochemical cell. The applied voltage produced an electric field which is opposed to the attraction forces between excited electrons and the holes that are produced during photocatalysis. This results in excited electrons being removed from the surface of TiO$_2$, minimizing the electron–hole recombination [39–41,73]. The same mesh was tested a second time, showing an efficiency drop of 14% compared to the initial run. Literature on the loss of degradation efficiency varies considerably. It was reported from no loss [41] to 7% [74] and 8% [26] efficiency reduction during the repeated cycles. The mechanism for the reduction in the photocatalytic activity after repeated use is the same as discussed for TiO$_2$ mesh earlier.

3.6. Comparison of Three Forms of the Photocatalyst and Electro-Photocatalysis

The required energy to achieve 60% degradation was used to compare the efficiency of the four studied photocatalysts. In the experiments using the mesh photocatalyst, the average value of the energy consumption of the four experiments is used for the purpose of comparison.

Energy (E_{lamp} in J/cm^3) was calculated based on the LED output in mW/cm^2 and the surface area of the photocatalyst. Therefore, the variation of the surface area of the different photocatalysts was taken into account. The degradation efficiency or the normalized degraded mass of 2,4-D (ΔM) was calculated based on the mass difference during the photocatalytic reaction using the following equation where M_0 is the initial mass of 2,4-D in the solution, and M_t is its mass at various sampling intervals:

$$\Delta M = \frac{M_0 - M_t}{M_0}, \tag{2}$$

In electro-photocatalysis, the applied excessive voltage was accounted in energy calculations. The following equations were used to calculate the total consumed energy per volume of the sample (E_{total} in J/cm^3):

$$E_{total} = E_{lamp} + E_{electro-chemical\ cell}, \tag{3}$$

$$E_{electro-chemical\ cell} = \frac{V \times I \times t}{V_s} \tag{4}$$

where V is the average potential of the electrochemical cell (in V), I is the average of measured current during irradiation (in A), t is time (in s) and V_s is the volume of the solution (in cm^3) [75,76]. The results are summarized in Table 3, and illustrated in Figure 6.

Table 3. Experimental results of 2,4-D degradation in various photocatalyst substrates.

Photocatalyst	Surface Area of the Photocatalyst (cm^2)	Energy (J/cm^3)	$T_{1/2}$ (h)	K Value ($mg\ L^{-1}\ h^{-1}$)	R^2
Anodized plate	4.97	114.5	7.97	3.07	0.99
Anodized mesh	2.76	80.3	7.09	3.45	0.99
Electro-photocatalysis	2.76	112.2	4.86	5.04	0.99
Photospheres	12.56	80.3	5.48	4.55	0.99

It was observed that energy consumption was at the same level in the anodized plate (i.e., 114.5 J/cm^3) and electro-photocatalysis (i.e., 112.2 J/cm^3). Energy consumption was also similar in the photospheres experiment at 80.3 J/cm^3, with the average energy consumption of the four studied anodized mesh at 80.3 J/cm^3. In comparison to the photospheres experiment and the mesh, energy consumption was 39% higher on average when the anodized plate and electro-photocatalysis were used.

The lower efficiency and higher energy consumption of the anodized plate is due to the 2D geometry of its planar surface compared to the 3D nanostructures on the mesh [26]. Although the dimensions of the plate and mesh were similar, void spaces in the mesh resulted in a significantly lower surface area available for light absorbance. In contrast, the three-dimensional surface morphology of TiO_2 nanotubes grown on a mesh are more efficient in the absorbance of the scattered light compared to the 2D geometry of the planar surface of the titanium plate despite having a lower surface available for irradiation. Moreover, the interstitial fissures between the nanotubes allow contaminant molecules easier access to the photocatalyst surface area, which enhances the photocatalytic activity.

This phenomenon has been studied by other researchers [26] in the photocatalytic degradation of methyl orange under a high-pressure mercury lamp with the wavelength of 365 nm. It was demonstrated that under similar conditions and equal time (360 min), the photocatalytic degradation of methyl orange was 18% higher when an anodized mesh was compared with an anodized plate.

The outperformance of the photospheres in comparison to the plate can be attributed to the increased access to the catalyst surface area by photons and the reactants—a primary factor in determining the degradation rate in photocatalysis. Other studies have also demonstrated similar results [51].

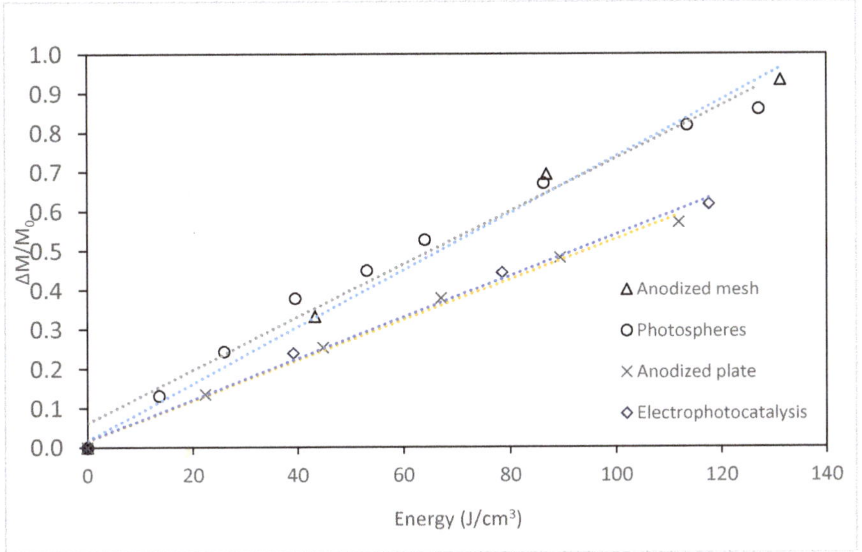

Figure 6. Comparison between photocatalysts based on the mass difference versus energy.

It was observed that with a similar amount of energy, the degradation efficiency of the anodized mesh was comparable with photospheres. This is due to the improved photocatalytic activity of the supported TNTs and 3D structures of TiO_2 surrounding the mesh wires, compared to anatase TiO_2, mounted on the glass spheres. Moreover, it has significant advantages because a separation step after treatment is not required, and it can be removed easily when the treatment is completed.

Albeit electro-photocatalysis resulted in the lowest half-life time (i.e., 4.86 h) compared to the methods studied here, but its energy consumption was at the same level as the anodized plate. This outcome was also in-line with other published comparative studies [41,42,75–77]. The enhanced efficiency is attributed to the presence of the electrical bias between the electrodes which minimized the electron-hole recombination on the photocatalyst and increased the quantum yield of the photocatalytic degradation process. The primary disadvantage of this system is the required additional electricity to the photoanode as well as the increased complexity of the process, which leads to a higher capital and operational costs [75].

The studied semi-passive system using the anodized plate is also compared with a flow-through photoreactor, using the same anodized plate and the same light source to degrade 2,4-D in a Killex® solution [49]. During 2 h of irradiation, the photoreactor resulted in 78% higher degradation efficiency of ΔM = 0.25 compared to ΔM = 0.14 in the semi-passive system. Energy consumption of the semi-passive system was 44.8 J/cm³ at ΔM = 0.25 which was 3.6 times higher than the photoreactor (12.5 J/cm³). This difference is due to the slower kinetics of the photocatalytic degradation in the semi-passive system due to the lack of mixing that minimizes mass transfer. The reported pseudo first-order kinetic rate constant was 648 h^{-1} in the reactor, compared to the zero-order rate constant of 3.07 mg L^{-1} h^{-1} in the semi-passive photocatalysis. It can be concluded that the semi-passive photocatalysis consumes more energy and results in a lower degradation efficiency during a same period of time when compared to a photoreactor.

Although photospheres have recently been broadly and successfully studied for degradation of environmental contaminants [20,78–81], their usage is still constrained by problems with separation after treatment. Moreover, ensuring the even dispersion of the photospheres for a semi-passive treatment, with no mixing, may be a challenge in practical applications. Furthermore, a direct comparison between the publications in this field and the results presented in this paper is not feasible due to the different operational parameters and the variety of the studied contaminants. The feasibility of utilization of the robust anodized TiO_2 based photocatalysts, designing a system for field applications at a larger scale, and studying the system to degrade various organic contaminants in water will be the direction for future research.

4. Conclusions

- The feasibility of using four types of TiO_2 photocatalysts under UV-LED irradiation (i.e., $\lambda = 365$ nm) in a semi-passive system was investigated.

- Energy consumption (J/cm^3) was used to rank the degradation efficiency of the photocatalysts. Photospheres (80.3 J/cm^3) and anodized titanium mesh (80.3 J/cm^3) showed similar efficiencies followed by electro-photocatalysis (112.2 J/cm^3) and the anodized plate (114.5 J/cm^3).

- Although the electro-photocatalysis rate of reaction was 64% higher than the photocatalysis with anodized plates, and 46% higher than the photocatalysis using the anodized mesh, it required additional energy and control during the photocatalytic degradation process.

- Increasing the loadings of the photospheres enhanced the kinetics of the degradation reaction from 4.12 mg L^{-1} h^{-1} to 4.55 mg L^{-1} h^{-1}, but further increases in the loadings reduced the rate of reaction to 4.36 mg L^{-1} h^{-1}. The variation in the rate of reaction, validated the deactivation of the originally activated species of TiO_2 due to the collision mechanism and the UV screening effect.

- Though the degradation rate and efficiency was high with photospheres, they still would require a separation step after treatment. This minimizes their attractiveness for application under ambient environments.

- Studying the depth of the photocatalyst from the surface showed at the range of 1 cm to 4 cm, the effect of the mass of the contaminant in the solution supersedes the effect of the light penetration.

- Anodized mesh showed 29% higher efficiency with 56% surface area of the anodized plate, but its performance varied during repetitive usage. The resulted variation is associated with the ability of the anodization process to generate a uniform mesh with a stable photocatalyst on its surface. This emphasizes the importance of improving the anodization process to produce a robust and uniform mesh which will be considered in future studies.

Supplementary Materials: The following are available online at http://www.mdpi.com/2073-4441/11/3/621/s1, Figure S1. XRD of titanium dioxide mesh, Figure S2. XRD of titanium dioxide plate, Figure S3. XRD of titanium dioxide photospheres.

Author Contributions: G.H. designed the study and conducted the experiments, analyzed data and wrote the original draft of the paper; J.H. edited the paper and supported data analysis; G.A. oversaw the research and contributed in experimental design, data analysis and reviewed the final draft; Late C.H.L. oversaw the research and experimental design. He could not review this manuscript.

Funding: This research was partially funded by CMC Research Institutes, MITACS and Natural Sciences and Engineering Research Council of Canada (NSERC).

Acknowledgments: The authors would like to thank Mitacs, and CMC Research Institutes and Natural Sciences and Engineering Research Council of Canada (NSERC) for providing partial funding for the project.

Conflicts of Interest: The authors declare no conflict of interest. The funders had no role in the design of the study; in the collection, analyses, or interpretation of data; in the writing of the manuscript, and in the decision to publish the results.

References

1. Fujishima, A.; Honda, K. Electrochemical evidence for the mechanism of the primary state of photosynthesis. *Bull. Chem. Soc. Jpn.* **1971**, *44*, 1148–1150. [CrossRef]

2.	Mondal, K. Recent advances in the synthesis of metal oxide nanofibers and their environmental remediation applications. *Inventions* **2017**, *2*, 9. [CrossRef]

3.	Chong, M.N.; Jin, B.; Chow, C.W.K.; Saint, C. Recent developments in photocatalytic water treatment technology: A review. *Water Res.* **2010**, *44*, 2997–3027. [CrossRef]

4.	Wang, C.; Liu, H.; Qu, Y. TiO$_2$-based photocatalytic process for purification of polluted water: Bridging fundamentals to applications. *J. Nanomater.* **2013**, *2013*, 319637. [CrossRef]

5.	Toor, A.P.; Verma, A.; Jotshi, C.K.; Bajpai, P.K.; Singh, V. Photocatalytic degradation of 3, 4-dichlorophenol using TiO$_2$ in a shallow pond slurry reactor. *Indian J. Chem. Technol.* **2005**, *12*, 75–81.

6.	Yu, L.; Achari, G.; Langford, C.H. LED-based photocatalytic treatment of pesticides and chlorophenols. *J. Environ. Eng.* **2013**, *139*, 1146–1151. [CrossRef]

7.	Ikehata, K.; El-Din, M.G.; Snyder, S.A. Ozonation and advanced oxidation treatment of emerging organic pollutants in water and wastewater. *Ozone Sci. Eng.* **2008**, *30*, 21–26. [CrossRef]

8.	Augugliaro, V.; Bellardita, M.; Loddo, V.; Palmisano, G.; Palmisano, L.; Yurdakal, S. Overview on oxidation mechanisms of organic compounds by TiO$_2$ in heterogeneous photocatalysis. *J. Photochem. Photobiol. C Photochem. Rev.* **2012**, *13*, 224–245. [CrossRef]

9.	Pelaez, M.; Nolan, N.T.; Pillai, S.C.; Seery, M.K.; Falaras, P.; Kontos, A.G.; Dunlop, P.S.M.; Hamilton, J.W.J.; Byrne, J.A.; O'Shea, K.; et al. A review on the visible light active titanium dioxide photocatalysts for environmental applications. *Appl. Catal. B Environ.* **2012**, *125*, 331–349. [CrossRef]

10.	Blake, D.M. Bibliography of work on the heterogeneous photocatalytic removal of hazardous compounds from water and air. *Natl. Renew. Energy Lab.* **2001**, *4*, 1–265. [CrossRef]

11.	Zhang, B.; Cao, S.; Du, M.; Ye, X.; Wang, Y.; Ye, J. Titanium Dioxide (TiO$_2$) Mesocrystals: Synthesis, Growth Mechanisms and Photocatalytic Properties. *Catalysis* **2019**, *9*, 91. [CrossRef]

12.	Fagan, R.; McCormack, D.E.; Dionysiou, D.D.; Pillai, S.C. A review of solar and visible light active TiO$_2$ photocatalysis for treating bacteria, cyanotoxins and contaminants of emerging concern. *Mater. Sci. Semicond. Process.* **2015**, *42*, 2–14. [CrossRef]

13.	Lavand, A.B.; Malghe, Y.S. Visible light photocatalytic degradation of 4-chlorophenol using C/ZnO/CdS nanocomposite. *J. Saudi Chem. Soc.* **2015**, *19*, 471–478. [CrossRef]

14.	Huang, Z.F.; Song, J.; Pan, L.; Zhang, X.; Wang, L.; Zou, J.J. Tungsten oxides for photocatalysis, electrochemistry, and phototherapy. *Adv. Mater.* **2015**, *27*, 5309–5327. [CrossRef] [PubMed]

15.	Zheng, H.; Ou, J.Z.; Strano, M.S.; Kaner, R.B.; Mitchell, A.; Kalantar-Zadeh, K. Nanostructured tungsten oxide–properties, synthesis, and applications. *Adv. Funct. Mater.* **2011**, *21*, 2175–2196. [CrossRef]

16.	Smith, Y.R.; Sarma, B.; Mohanty, S.K.; Misra, M. Light-assisted anodized TiO$_2$ nanotube arrays. *ACS Appl. Mater. Interfaces* **2012**, *4*, 5883–5890. [CrossRef] [PubMed]

17.	Heller, A.; Brock, J. Materials and Methods for Photocatalyzing Oxidation of Organic Compounds on Water. U.S. Patent 4,997,576, 5 March 1991. Available online: http://www.freepatentsonline.com/4997576.html (accessed on 26 March 2019).

18.	Gjipalaj, J.; Alessandri, I. Easy recovery, mechanical stability, enhanced adsorption capacity and recyclability of alginate-based TiO$_2$ macrobead photocatalysts for water treatment. *J. Environ. Chem. Eng.* **2017**, *5*, 1763–1770. [CrossRef]

19.	Robert, D.; Keller, V.; Keller, N. Immobilization of a semi-conductor photocatalyst on solid supports, methods, materials and applications. In *Photocatalysis and Water Purification: From Fundamentals to Recent Applications*; Lu, M., Pichat, P., Eds.; John Wiley & Sons, Incorporated: Hoboken, NJ, USA, 2013; pp. 145–172. ISBN 9783527645411.

20.	Abdel-Maksoud, Y.; Imam, E.; Ramadan, A. TiO$_2$ solar photocatalytic reactor systems: Selection of reactor design for scale-up and commercialization—Analytical review. *Catalysts* **2016**, *6*, 138. [CrossRef]

21.	Sakthivel, S.; Shankar, M.V.; Palanichamy, M.; Arabindoo, B.; Murugesan, V. Photocatalytic decomposition of leather dye comparative study of TiO$_2$ supported on alumina and glass beads. *J. Photochem. Photobiol. A Chem.* **2002**, *148*, 153–159. [CrossRef]

22.	Sirisuk, A.; Hill, C.G.; Anderson, M.A. Photocatalytic degradation of ethylene over thin films of titania supported on glass rings. *Catal. Today* **1999**, *54*, 159–164. [CrossRef]

23.	Portjanskaja, E.; Krichevskaya, M.; Preis, S.; Kallas, J. Photocatalytic oxidation of humic substances with TiO$_2$-coated glass micro-spheres. *Environ. Chem. Lett.* **2004**, *2*, 123–127. [CrossRef]

24. Ge, M.; Cao, C.; Huang, J.; Li, S.; Chen, Z.; Zhang, K.-Q.; Al-Deyab, S.S.; Lai, Y. A review of one-dimensional TiO_2 nanostructured materials for environmental and energy applications. *J. Mater. Chem. A* **2016**, *4*, 6772–6801. [CrossRef]

25. Zeng, Q.; Xi, M.; Xu, W.; Li, X. Preparation of titanium dioxide nanotube arrays on titanium mesh by anodization in $(NH4)_2SO_4/NH_4F$ electrolyte. *Mater. Corros.* **2013**, *64*, 1001–1006. [CrossRef]

26. Liao, J.; Lin, S.; Zhang, L.; Pan, N.; Cao, X.; Li, J. Photocatalytic degradation of methyl orange using a TiO_2/Ti mesh electrode with 3D nanotube arrays. *ACS Appl. Mater. Interfaces* **2012**, *4*, 171–177. [CrossRef]

27. Roy, P.; Berger, S.; Schmuki, P. TiO_2 nanotubes: Synthesis and applications. *Angew. Chem.–Int. Ed.* **2011**, *50*, 2904–2939. [CrossRef]

28. Jun, Y.; Park, J.H.; Kang, M.G. The preparation of highly ordered TiO_2 nanotube arrays by an anodization method and their applications. *Chem. Commun.* **2012**, *48*, 6456–6471. [CrossRef]

29. Albu, S.P.; Ghicov, A.; Macak, J.M.; Hahn, R.; Schmuki, P. Self-organized, free-standing TiO_2 nanotube membrane for flow-through photocatalytic applications. *Nano Lett.* **2007**, *7*, 1286–1289. [CrossRef]

30. Motola, M.; Satrapinskyy, L.; Roch, T.; Šubrt, J.; Kupčík, J.; Klementová, M.; Jakubičková, M.; Peterka, F.; Plesch, G. Anatase TiO_2 nanotube arrays and titania films on titanium mesh for photocatalytic NOXremoval and water cleaning. *Catal. Today* **2017**, *287*, 59–64. [CrossRef]

31. Zhong, M.; Zhang, G.; Yang, X. Preparation of Ti mesh supported WO_3/TiO_2 nanotubes composite and its application for photocatalytic degradation under visible light. *Mater. Lett.* **2015**, *145*, 216–218. [CrossRef]

32. Cunha, D.L.; Kuznetsov, A.; Achete, C.A.; da Hora Machado, A.E.; Marques, M. Immobilized TiO_2 on glass spheres applied to heterogeneous photocatalysis: Photoactivity, leaching and regeneration process. *PeerJ.* **2018**, *4464*, 1–19. [CrossRef] [PubMed]

33. Hartley, A.C.; Moss, J.B.; Keesling, K.J.; Moore, N.J.; Glover, J.D.; Boyd, J.E. PMMA-titania floating macrospheres for the photocatalytic remediation of agro-pharmaceutical wastewater. *Water Sci. Technol.* **2017**, *75*, 1362–1369. [CrossRef]

34. Xing, Z.; Li, J.; Wang, Q.; Zhou, W.; Tian, G.; Pan, K.; Tian, C.; Zou, J.; Fu, H. A floating porous crystalline TiO_2 ceramic with enhanced photocatalytic performance for wastewater decontamination. *Eur. J. Inorg. Chem.* **2013**, 2411–2417. [CrossRef]

35. Magalhães, F.; Moura, F.C.C.; Lago, R.M. TiO_2/LDPE composites: A new floating photocatalyst for solar degradation of organic contaminants. *Desalination* **2011**, *276*, 266–271. [CrossRef]

36. Magalhães, F.; Lago, R.M. Floating photocatalysts based on TiO_2 grafted on expanded polystyrene beads for the solar degradation of dyes. *Sol. Energy* **2009**, *83*, 1521–1526. [CrossRef]

37. Bahreini, Z.; Heydari, V.; Hekmat, A.N.; Taheri, M.; Vahid, B.; Moradkhannejhad, L. A comparative study of photocatalytic degradation and mineralisation of an azo dye using supported and suspended nano-TiO_2 under UV and sunlight irradiations. *Pigment Resin Technol.* **2016**, *45*, 119–125. [CrossRef]

38. Salinaro, A.; Emeline, A.V.; Zhao, J.; Hidaka, H.; Ryabchuk, V.K.; Serpone, N. Terminology, relative photonic efficiencies and quantum yields in heterogeneous photocatalysis. Part I: Suggested protocol. *Pure Appl. Chem.* **1999**, *71*, 303–320. [CrossRef]

39. Turolla, A. Heterogeneous Photocatalysis and Electro-Photocatalysis on Nanostructured Titanium Dioxide for Water and Wastewater Treatment: Process Assessment, Modelling and Optimization. Ph.D. Thesis, Polytechnic University of Milan, Milan, Italy, 2014.

40. Zlamal, M.; Macak, J.M.; Schumulu, P.; Josef, K. Electrochemically assisted photocatalysis on self-organized TiO_2 nanotubes. *Electrochem. Commun.* **2007**, *9*, 2822–2826. [CrossRef]

41. Turolla, A.; Fumagalli, M.; Bestetti, M.; Antonelli, M. Electro-photocatalytic decolorization of an azo dye on TiO_2 self-organized nanotubes in a laboratory scale reactor. *Desalination* **2012**, *285*, 377–382. [CrossRef]

42. Wu, T.N.; Pan, T.C.; Chen, L.C. Electro-photocatalysis of aqueous methyl tert-butyl ether on a titanium dioxide coated electrode. *Electrochim. Acta* **2012**, *86*, 170–176. [CrossRef]

43. Shen, Y.; Li, F.; Li, S.; Liu, D.; Fan, L.; Zhang, Y. Electrochemically enhanced photocatalytic degradation of organic pollutant on β-PbO_2-TNT/Ti/TNT bifunctional electrode. *Int. J. Electrochem. Sci.* **2012**, *7*, 8702–8712.

44. Krieger, R. (Ed.) *Hayes' Handbook of Pesticide Toxicology*; Elsevier Science & Technology: Saint Louis, MO, USA, 2010; ISBN 9780080922010.

45. Health Canada Special Review of 2,4-D: Proposed Decision for Consultation. Available online: https://www.canada.ca/en/health-canada/services/consumer-product-safety/pesticides-pest-management/public/consultations/re-evaluation-note/2016/special-review-2-4-d/document.html#s2 (accessed on 8 February 2017).

46. Guillard, C.; Amalric, L.; D'Oliveira, J.C.; Delpart, H.; Hoang-Van, C.; Pichat, P. Heterogeneous photocatalysis: Use in water treatment and involvement in atmopheric chemistry. In *Aquatic and Surface Photochemistry*; Helz, G.R., Zepp, R.G., Crosby, D.G., Eds.; Lewis Publishers: Boca Raton, FL, USA, 1994; pp. 369–386. ISBN 0873718712.

47. Meunier, L.; Pilichowski, J.F.; Boule, P. Photochemical behaviour of 1,4-dichlorobenzene in aqueous solution. *Can. J. Chem. Can. Chim.* **2001**, *79*, 1179–1186. [CrossRef]

48. Heydari, G. Passive or Semi-Passive Photocatalytic Treatment of Organic Pollutants in Water. Master's Thesis, University of Calgary, Calgary, AB, Canada, 2018.

49. Radwan, E.K.; Yu, L.; Achari, G.; Langford, C.H. Photocatalytic ozonation of pesticides in a fixed bed flow through UVA-LED photoreactor. *Environ. Sci. Pollut. Res.* **2016**, *23*, 21313–21318. [CrossRef] [PubMed]

50. Eskandarian, M.R.; Choi, H.; Fazli, M.; Rasoulifard, M.H. Effect of UV-LED wavelengths on direct photolytic and TiO$_2$ photocatalytic degradation of emerging contaminants in water. *Chem. Eng. J.* **2016**, *300*, 414–422. [CrossRef]

51. Yu, L. *Light Emitting Diode Based Photochemical Treatment of Contaminants in Aqueous Phase*; University of Calgary: Calgary, AB, Canada, 2014.

52. Casterjon-Sanchez, V.H.; Lopez, R.; Ramon-Gonzalez, M.; Enriquez-Perez, A.; Camasho-Lopez, M.; Villa-Sanchez, G. Annealing Control on the Anatase/Rutile Ratio of Nanostructured Titanium Dioxide Obtained by Sol-Gel. *Crystals* **2018**, *9*, 22. [CrossRef]

53. Topalov, A.S.; Sojic, D.V.; Molnar-Gabor, D.A.; Abramovic, B.F.; Comor, M.I. Photocatalytic activity of synthesized nanosized TiO$_2$ towards the degradation of herbicide mecoprop. *Appl. Catal. B Environ.* **2004**, *54*, 125–133. [CrossRef]

54. Aguer, J.; Blachère, F.; Boule, P.; Garaudee, S.; Guillard, C. Photolysis of dicamba (3,6-dichloro-2-methoxybenzoic acid) in aqueous solution and dispersed on solid supports. *Int. J. Photoenergy* **2000**, *2*, 81–86. [CrossRef]

55. Fogarty, A.M.; Traina, S.J.; Tuovinen, O.H. Determination of Dicamba by Reverse-Phase HPLC. *J. Liq. Chromatogr.* **1994**, *17*, 2667–2674. [CrossRef]

56. Castellote, M.; Bengtsson, N. Principles of TiO$_2$ photocatalysis. In *Applications of Titanium Dioxide Photocatalysis to Construction Materials: State-of-the-Art Report of the RILEM Technical Committee 194-TDP*; Ohama, Y., Van Gemert, D., Eds.; Springer: Dordrecht, The Netherland, 2011; pp. 5–9. ISBN 9789400712966.

57. Li, L.; Wang, M. Advanced nanomaterials for solar photocatalysis. In *Advanced Catalytic Materials–Photocatalysis and Other Current Trends*; INTECH: London, UK, 2016; pp. 169–230. ISBN 9789535122449.

58. Calza, P.; Sakkas, V.; Medana, C.; Baiocchi, C.; Dimou, A.; Pelizzetti, E.; Albanis, T. Photocatalytic degradation study of diclofenac over aqueous TiO$_2$ suspensions. *Appl. Catal. B Environ.* **2006**, *67*, 197–205. [CrossRef]

59. Sarkar, S.; Das, R.; Choi, H.; Bhattacharjee, C. Involvement of process parameters and various modes of application of TiO$_2$ nanoparticles in heterogeneous photocatalysis of pharmaceutical wastes—A short review. *RSC Adv.* **2014**, *4*, 57250–57266. [CrossRef]

60. Hu, L.; Flanders, P.M.; Miller, P.L.; Strathmann, T.J. Oxidation of sulfamethoxazole and related antimicrobial agents by TiO$_2$ photocatalysis. *Water Res.* **2007**, *41*, 2612–2626. [CrossRef] [PubMed]

61. Neppolian, B.; Choi, H.C.; Sakthivel, S.; Arabindoo, B.; Murugesan, V. Solar/UV-induced photocatalytic degradation of three commercial textile dyes. *J. Hazard. Mater.* **2002**, *89*, 303–317. [CrossRef]

62. Yang, L.; Yu, L.E.; Ray, M.B. Degradation of paracetamol in aqueous solutions by TiO$_2$ photocatalysis. *Water Res.* **2008**, *42*, 3480–3488. [CrossRef] [PubMed]

63. Tsydenova, O.; Batoev, V.; Batoeva, A. Solar-enhanced advanced oxidation processes for water treatment: Simultaneous removal of pathogens and chemical pollutants. *Int. J. Environ. Res. Public Health* **2015**, *12*, 9542–9561. [CrossRef]

64. Gaya, U.I.; Abdullah, A.H. Heterogeneous photocatalytic degradation of organic contaminants over titanium dioxide: A review of fundamentals, progress and problems. *J. Photochem. Photobiol. C Photochem. Rev.* **2008**, *9*, 1–12. [CrossRef]

65. Lee, K.; Kim, D.; Roy, P.; Paramasivam, I.; Birajdar, B.I.; Spiecker, E.; Schmuki, P. Anodic formation of thick anatase TiO_2 mesosponge layers for high efficiency photocatalysis. *Mater. Sci.* **2010**, *7*, 1–10. [CrossRef]

66. Schmidt-Stein, F.; Thiemann, S.; Berger, S.; Hahn, R.; Schmuki, P. Mechanical properties of anatase and semi-metallic TiO_2 nanotubes. *Acta Mater.* **2010**, *58*, 6317–6323. [CrossRef]

67. Minagar, S.; Berndt, C.C.; Wang, J.; Ivanova, E.; Wen, C. A review of the application of anodization for the fabrication of nanotubes on metal implant surfaces. *Acta Biomater.* **2012**, *8*, 2875–2888. [CrossRef] [PubMed]

68. Wang, J.; Lin, Z. Anodic formation of ordered TiO_2 nanotube arrays: Effects of electrolyte temperature and anodization potential. *J. Phys. Chem. C* **2009**, *113*, 4026–4030. [CrossRef]

69. Yu, J.; Wang, B. Effect of calcination temperature on morphology and photoelectrochemical properties of anodized titanium dioxide nanotube arrays. *Appl. Catal. B Environ.* **2010**, *94*, 295–302. [CrossRef]

70. Jarosz, M.; Kapusta-Kołodziej, J.; Jaskuła, M.; Sulka, G.D. Effect of different polishing methods on anodic titanium dioxide formation. *J. Nanomater.* **2015**, *2015*, 295126. [CrossRef]

71. Fujishima, A.; Rao, T.N.; Tryk, D.A. Titanium dioxide photocatalysis. *J. Photochem. Photobiol. C Photochem. Rev.* **2000**, *1*, 1–21. [CrossRef]

72. Chekir, N.; Boukendakdji, H.; Igoud, S.; Taane, W. Solar energy for the benefit of water treatment: Solar photoreactor. *Procedia Eng.* **2012**, *33*, 174–180. [CrossRef]

73. Mackak, J.; Tsuchiya, H.; Ghicov, A.; Yasuda, K.; Hahn, R.; Bauer, S.; Scumula, P. TiO_2 nanotubes: Self-organized electrochemical formation, properties and applications. *Curr. Opin. Solid State Mater. Sci.* **2007**, *11*, 3–18. [CrossRef]

74. Eskandarloo, H.; Hashempour, M.; Vicenzo, A.; Franz, S.; Badiei, A.; Behnajady, M.A.; Bestetti, M. High-temperature stable anatase-type TiO_2 nanotube arrays: A study of the structure-activity relationship. *Appl. Catal. B Environ.* **2016**, *185*, 119–132. [CrossRef]

75. Moreira, F.C.; Boaventura, R.A.R.; Brillas, E.; Vilar, V.J.P. Electrochemical advanced oxidation processes: A review on their application to synthetic and real wastewaters. *Appl. Catal. B Environ.* **2017**, *202*, 217–261. [CrossRef]

76. Martínez-Huitle, C.A.; Brillas, E. Decontamination of wastewaters containing synthetic organic dyes by electrochemical methods: A general review. *Appl. Catal. B Environ.* **2009**, *87*, 105–145. [CrossRef]

77. Li, X.Z.; Liu, H.L.; Li, F.B.; Mak, C.L. Photoelectrocatalytic oxidation of Rhodamine B in aqueous soultion using Ti/TiO_2 mesh photoelectrodes. *J. Environ. Sci. Heal. Part A* **2007**, *37*, 55–69. [CrossRef]

78. Leshuk, T.; Krishnakumar, H.; de Oliveira Livera, D.; Gu, F. Floating photocatalysts for passive solar degradation of naphthenic acids in oil sands process-affected water. *Water* **2018**, *10*, 202. [CrossRef]

79. Wang, J.; He, B.; Kong, X.Z. A study on the preparation of floating photocatalyst supported by hollow TiO_2 and its performance. *Appl. Surf. Sci.* **2015**, *327*, 406–412. [CrossRef]

80. Yuan, J.; An, Z.-G.; Zhang, J.-J.; Li, B. Synthesis and properties of hollow glass spheres/TiO_2 composite. *Imaging Sci. Photochem.* **2012**, *30*, 447–455.

81. Miranda-García, N.; Suárez, S.; Sánchez, B.; Coronado, J.M.; Malato, S.; Maldonado, M.I. Photocatalytic degradation of emerging contaminants in municipal wastewater treatment plant effluents using immobilized TiO_2 in a solar pilot plant. *Appl. Catal. B Environ.* **2011**, *103*, 294–301. [CrossRef]

Article

An Amphiphilic, Graphitic Buckypaper Capturing Enzyme Biomolecules from Water

Shahin Homaeigohar * and Mady Elbahri *

Nanochemistry and Nanoengineering, School of Chemical Engineering, Department of Chemistry and Materials Science, Aalto University, Kemistintie 1, 00076 Aalto, Finland
* Correspondence: shahin.homaeigohar@aalto.fi (S.H.); mady.elbahri@aalto.fi (M.E.); Tel.: +358-50-431-9831

Received: 13 November 2018; Accepted: 18 December 2018; Published: 20 December 2018

Abstract: The development of carbon nanomaterials for adsorption based removal of organic pollutants from water is a progressive research subject. In this regard, carbon nanomaterials with bifunctionality towards polar and non-polar or even amphiphilic undesired materials are indeed attractive for further study and implementation. Here, we created carbon buckypaper adsorbents comprising amphiphilic (oxygenated amorphous carbon (a-CO$_x$)/graphite (G)) nanofilaments that can dynamically adsorb organic biomolecules (i.e., urease enzyme) and thus purify the wastewaters of relevant industries. Given the dynamic conditions of the test, the adsorbent was highly efficient in adsorption of the enzyme (88%) while being permeable to water (4750 L·h^{-1}m^{-2}bar^{-1}); thus, it holds great promise for further development and upscaling. A subsequent citric acid functionalization declined selectivity of the membrane to urease, implying that the biomolecules adsorb mostly via graphitic domains rather than oxidized, polar amorphous carbon ones.

Keywords: carbon; nanofiber; membrane; urease; biomolecules; water treatment

---

1. Introduction

As a global challenge, water scarcity is expanding to major parts of the world, threatening human beings' lives. This crisis can have different origins but undoubtedly water pollution from industry and from urban communities is a main one. Amongst the variety of water pollutants, the organic ones such as proteins and biomolecules play a determining role. These substances even at a negligible amount, <1% of the entire contamination in a river, for instance, can deplete the oxygen present in water and cause the death of living creatures in that ecosystem [1]. Water recycling via purification can somewhat remediate this problem but necessitates the development of advanced water treatment systems. Micro-, ultra- and nanofiltration membranes are typically utilized for wastewater treatment. Their purification action mainly relies on sieving of the pollutants, and thus they require a porous structure whose pore size is less than the size of the solute to be separated. Other than the membranes, in a sustainable manner and using conventional and also emerging materials, functionalized adsorbents have shown applicability in the removal of even molecules and tiny pollutants based on physical/chemical interactions or biological functions [2–7]. Accordingly, there is no need for the construction of porous materials with small pore sizes that could impose high pressure differences. Moreover, a functionalized adsorbent with a surface decorated by particular functional groups can discriminate or entrap molecules in a selective manner [8].

Electrospun nanofibrous adsorbents have shown promising capabilities for selective water remediation. Their structure possesses a high interconnected porosity and huge surface area that in case of functionalization can efficiently separate functional pollutants, e.g., ions, dye molecules, organics, etc. While the high porosity realizes a significant permeability and with that, energy efficiency, the expansive surface area enables the notable functionalization necessary for highly selective adsorbents. In this regard, several biofunctionalized nanofibrous membranes made of polyurethane, polysulfone, polyacrylonitrile, and cellulose have been tested for the separation of protein and enzyme (e.g., Immunoglobulin G (IgG), Bovine Serum Albumin (BSA), lipase, bromelain, etc.) [8–11]. In our studies [2,4,12], we also developed a biofunctionalized nanofibrous adsorbent composed of Bovine Serum Albumin and poly(acrylonitrile-co-glycidyl methacrylate) (PANGMA), as the functional agent and polymer nanofiber, respectively, that could offer a significant metal nanoparticle and biomolecule removal efficiency while being highly water permeable. This adsorbent was synthesized in a simple fashion versus the other previously developed systems [8,13]. The separation tests were performed under the most tricky conditions, i.e., dynamically and with a low protein amount (a few $mg \cdot L^{-1}$ instead of $mg \cdot mL^{-1}$ adopted by References [8,10,11,14]) and with a size scale of pollutants, potentially passing readily through a macroporous nanofibrous structure. Despite such circumstances, the adsorbent was successful in the removal of nanoparticles (97%) as well as proteins (88% BSA and 81% *Candida antarctica* Lipase B (Cal-B)). In another research, we developed a nanofibrous adsorbent comprising polyethersulfone (PES) nanofibers that were functionalized by the inclusion of vanadium oxide (V_2O_5) nanoparticles [6]. This adsorbent system was able to separate methylene blue (MB) dye from water with an efficiency of 85% under alkaline condition and high temperature.

Despite the various merits of the above-mentioned systems in the adsorption of diverse water pollutants, their synthesis and functionalization are not one pot. As a step forward to meet this need, recently, we developed carbon buckypapers based on amphiphilic carbon nanofilaments [15]. The nanofilaments are composed of oxygenated amorphous carbon ($a\text{-}CO_x$) and graphite (G), and thus are able to adsorb both polar (e.g., dye) and non-polar (e.g., oil) water pollutants efficiently. Already investigating the applicability of the amphiphilic graphitic buckypaper in discrimination of polar and non-polar contaminants, here, we challenge the buckypaper adsorbents with an amphiphilic water pollutant. For this sake, biomolecules (i.e., urease enzyme), one of the major organic pollutants that can adversely affect the water ecosystems, will be considered.

2. Experimental

Materials: polyacrylonitrile (PAN) (200,000 $g \cdot mol^{-1}$, purity 99.5%) and dimethylformamide (DMF) (purity 99%) were purchased from Dolan GmbH (Kelheim, Germany) and Merck (Darmstadt, Germany), respectively. Urease enzyme (impurity; ammonium < 4–10 $\mu mol \cdot U^{-1}$ enzyme) and citric acid (citric acid monohydrate, ACS reagent $\geq$ 99.0%) were obtained from Sigma-Aldrich (Saint Louis, MO, USA). All the materials were used as received.

Synthesis: The precursor PAN nanofibers were synthesized by electrospinning. To do so, employing a syringe pump (Harvard Apparatus, Holliston, MA, USA), a solution of PAN (8 wt % in DMF) was fed steadily (1 $mL \cdot h^{-1}$) into a needle (0.8 mm inner diameter with a circular opening). Upon electrifying the solution with a voltage of 20 kV (Heinzinger Electronic GmbH, Rosenheim, Germany), PAN was electrospun on an aluminum foil. The as-synthesized PAN nanofibers underwent oxidative stabilization and were heated in air at 250 °C for 2 h within a furnace with maximum operational temperature of 1250 °C (Linn Elektro Therm). In the next step, the oxidized nanofibers were carbonized under argon atmosphere at 1250 °C for half an hour with a heating rate of 5 °C·min^{-1} and then cooled down to the room temperature at a same rate.

Due to the extreme brittleness of the graphitized nanofibers, challenging their handling as a freestanding membrane, they were suspended in distilled water (10 mL) and underwent an ultrasonication process for 2 min at a power of 20%. The $a\text{-}CO_x/G$ nanofibers under the influence of ultrasonication are disintegrated as suspended nanofilaments that can be subsequently cast on a circular poly(phenylene sulfide) (PPS) technical nonwoven (3.5 cm in diameter). As a control group, $a\text{-}CO_x/G$ nanofilaments were also functionalized by citric acid (CA). To do this, CA (30 $mg \cdot mL^{-1}$)) was added to the aqueous suspension to be ultrasonicated.

Characterization: The $a\text{-}CO_x/G$ nanofilaments were characterized in terms of morphology by scanning electron microscopy (SEM) (LEO 1550VP Gemini from Carl ZEISS, Jena, Germany) and an atomic force microscope (AFM) (MultiMode^{TM} Atomic Force Microscope from Bruker AXS, Madison, WI, USA). The surface chemistry of the a-CO$_x$/G nanofilaments was analyzed by FTIR (ALPHA (ATR-Ge, ATR-Di) from BRUKER Optik GmbH, Ettlingen, Germany). The a-CO$_x$/G buckypaper's pore size distribution was determined by an automated capillary flow porometer (Porous Materials Inc. (PMI), Ithaca, NY, USA).

The urease retention efficiency of the buckypapers was assessed using the corresponding aqueous solutions in a dead-end manner and by employing a lab-built set-up [16]. The set-up's reservoir contained 200 mL urease solution (1 g·L^{-1}) which permeated through the buckypapers under a 0.5 bar pressure. Based on a constructed standard urease calibration curve, the permeate's urease concentration was determined by UV-vis spectroscopy (HITACHI U3000, HITACHI, Tokyo, Japan). The urease retention efficiency (RE) was calculated via Equation (1):

$$\text{RE} = \left(1 - \frac{C_p}{C_f}\right) \times 100\% \tag{1}$$

where C_p and C_f represent the permeate's and feed's urease concentration, respectively. The permeation time was also recorded and the permeate permeance was calculated via Equation (2) [17]:

$$J = \frac{Q}{A \times \Delta t \times \Delta P} \tag{2}$$

where J is the permeate permeance (L·h^{-1}m^{-2}bar^{-1}), Q is the collected volume (L) of the permeate, A is the effective filtration area of the buckypapers (m^2), Δt is the collecting time (h), and ΔP is the pressure difference (bar). The permeance measurements were done for three 50 mL permeates to ascertain the consistency of the buckypapers' performance. It is worthy to note that considering the hydrophobic, large microfibers of the PPS support layer, providing huge pore sizes, no significant contact and interaction with the urease molecules passing through the carbon layer can be envisioned. Thus, only the buckypaper is responsible for the reported removal efficiency and permeance.

The electrical conductivity of the buckypapers as non-functionalized and CA-functionalized before and after urease adsorption was measured by a four-point probe test. At least five measurements were done on different parts of the buckypapers, and the error bars were calculated. The thickness of the samples to be considered in the conductivity measurement was already measured by a digital micrometer (Deltascope®MP2C from Fischer, Windsor, CT, USA).

3. Results and Discussion

The developed buckypaper consists of the a-CO$_x$/G nanofilaments, randomly arranged but with no sign of clustering. SEM image, as shown in Figure 1a, clearly verifies this fact and the preservation of a porous structure that guarantees optimum water permeability. Moreover, as seen in Figure 1b, the nanofilaments' tips are exposed to the surrounding medium, and thus, this raises the interactivity of the material with the biomolecule pollutants. In fact, the nanofilaments are able to capture urease through adsorption not only on their body, but also on their cross-sections. AFM images, as shown in Figure 1c, provide insight into the dimensions and morphology of the nanofilaments individually.

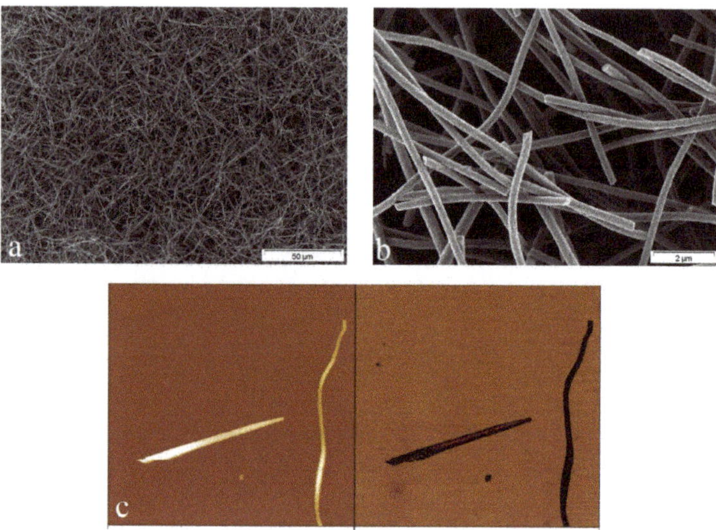

Figure 1. Scanning electron microscopy (SEM) images show morphology of the nanofilaments at (**a**) a low and (**b**) high magnification. (**c**) Atomic force microscope (AFM) micrographs imply the nanofilaments' dimensions and morphology.

Pore size measurement via a bubble point test, as shown in Figure 2, implies that the pore size lies in the submicron range as small as 700 nm. This pore size distribution qualifies the structure as a microfiltration (MF) membrane [18,19] that could hardly stop the passage of tiny organic pollutants, particularly under hydrodynamic pressure. It is worthy to note that the adsorption tests reported in the literature are typically performed in a static, batch mode that maximizes the contact time of the adsorbent and solutes. To the contrary, here, we adopted an adsorption test in a dynamic mode under a hydrodynamic pressure that could challenge the adsorption efficiency of our samples. This test was performed in three successive steps, in each, 50 mL of the solution was passed through. Such a style can highlight the repeatability of the result and also stress robust bonding of the urease molecules with the nanofilaments in case they are not released in the next steps and if efficiency does not decline.

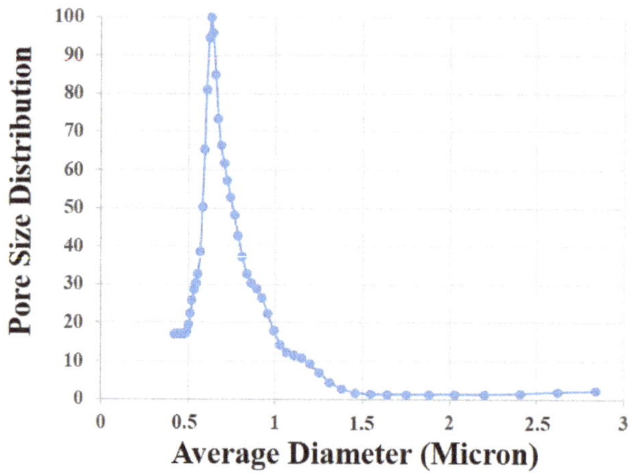

Figure 2. Pore size distribution of the a-CO_x/G buckypaper measured by a bubble point test.

Despite possessing pore sizes in the MF scale, the buckypaper showed a promising urease separation efficiency, as shown in Figure 3a. A removal efficiency of 88.5% was recorded after permeation of 150 mL urease aqueous solution. An ascending trend from 50 mL (75%) to 150 mL (88%) in urease removal efficiency is observed. As we previously proved [15], the nanofilaments have oxygen-based functional groups including carbonyl and hydroxyl that enable interaction, i.e., hydrogen bonding with amino acid units of urease. In addition to hydrogen bonding, the positively charged amine groups of urease and the negatively charged oxygen-based functional groups of a-CO$_x$ segments can electrostatically interact. On the other hand, major graphitic regions allow for π-π interaction with non-polar domains of urease. For a similar DNA–CNT system, van der waals forces have been introduced as an adsorption driving factor with a larger impact than hydrophobic forces [20]. For the urease molecules, several intramolecular bondings between different functional groups could also be envisaged. Accordingly, some molecules interact through their less polar and non-polar zones with the nanofilaments. Thus, collectively, different parts of the nanofilaments are able to adsorb urease molecules via interaction with their corresponding regions. This feature can stabilize the enzyme on its substrate and prevent its conformational change that can lead to loss of enzyme activity, which is beneficial for further application as, e.g., a biocatalyst [21–23]. Moreover, the huge surface area of the buckypaper minimizes the diffusion pathway for the reaction products, thus enhancing the efficiency of the immobilized enzyme. ATR-FTIR spectra, as shown in Figure 4a, clearly verify the adsorption of urease onto the nanofilaments. Before the adsorption, the strong peak located at 1589 cm^{-1} represents the unoxidized sp^2 C=C groups of the graphitic segments of the nanofilaments, which resulted from the aromatization process during the thermostabilization of PAN nanofibers [24,25]. The second evident groups at 1000–1300 (two bands) and 3800 cm^{-1} imply a C–OH bond [15]. After the adsorption, the main chemical bonds related to urease emerge on the nanofilaments. These bonds represented by ATR-FTIR characteristic peaks are marked in Figure 4a.

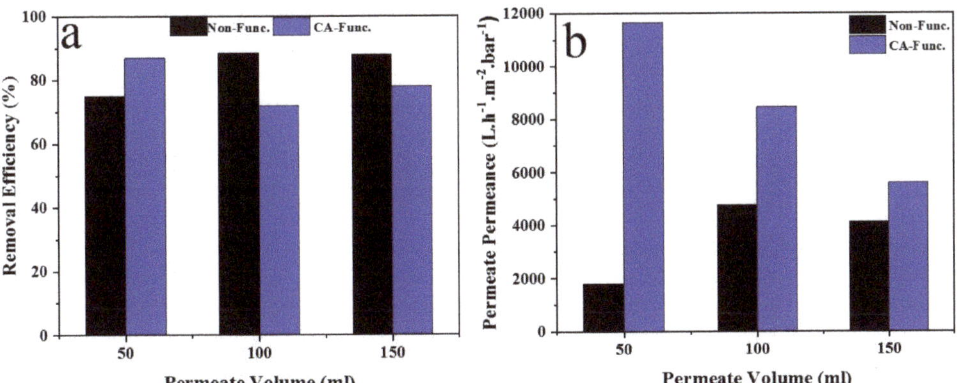

Figure 3. (**a**) Urease removal efficiency; (**b**) permeate permeance of the buckypapers in two classes of non-functionalized and CA-functionalized.

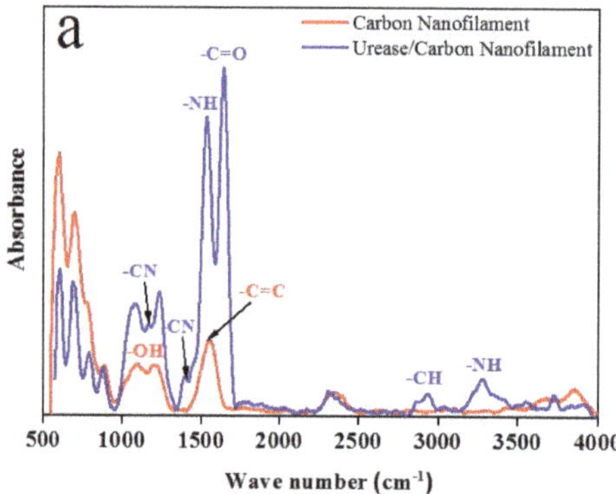
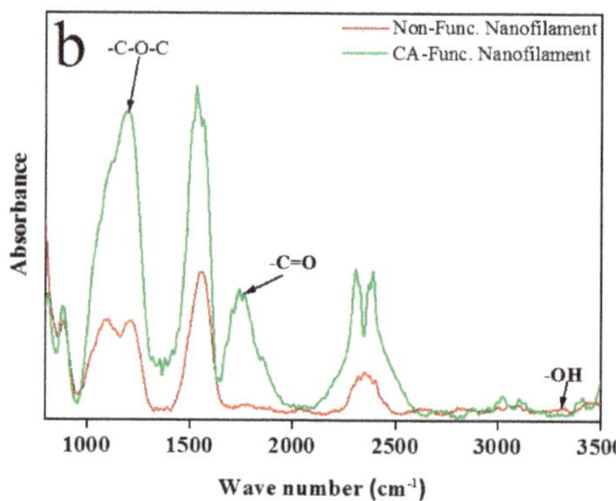

Figure 4. ATR-FTIR spectra compare the surface chemistry of carbon nanofilaments before and after (**a**) urease adsorption and (**b**) CA functionalization.

The increasing trend of urease removal efficiency shows that the adsorption of urease is robust and further passage of the solution does not result in its release into water. This enhancement of removal efficiency can be attributed to a strong intermolecular interaction between the adsorbed urease molecules and solutes via peptide–peptide interactions [26]. Interestingly, adsorption of urease molecules enhances permeate permeance of the buckypaper after the first round, due to a hydrophilization effect, as shown in Figure 3b. It is worthy to note that the pure water permeance of non- and CA-functionalized were measured as $\approx$8670 and 14000 (L·h^{-1}m^{-2}bar^{-1}), respectively, and adsorption of urease declines, most likely due to pore blockage and loss of porosity. In contrast to the non-functionalized samples, CA-functionalization and the emergence of various functional groups such as carboxyl and hydroxyl (Figure 4b) slightly lower the removal efficiency as far as the filtration is continued. While a high efficiency of 87% is seen at the onset of the experiment, it declines to 78% at 150 mL permeate volume. The reason could be found at less available binding sites for urease molecules or even the release of the previously adsorbed ones because of less graphitic regions that most likely have played a more important role in the stable adsorption of urease molecules rather than polar groups (hydrogen bonding or electrostatic interaction). However, still efficiency is as promising as 78%. It is worth nothing that permeate permeance for CA-buckypapers are significantly higher than that for the non-functionalized ones due to their hydrophilicity. The descending trend of permeate permeance in this class of adsorbents could be attributed to their declined hydrophilicity compared to the neat or fresh CA-functionalized samples due to the adsorption of the urease molecules. Slightly enhanced hydrophobicity along with the accumulation of the adsorbed molecules on the nanofilaments that lower pore size, cooperatively increase the resistance against water permeation.

The adsorption experiment performed here can be regarded as a proof of concept witnessing the applicability of the buckypaper adsorbent in the removal of urease molecules as a model for biomolecule pollutants from water. In this regard, taking into account the effect of environmental factors such as pH, temperature, ionic strength, adsorption time, and urease concentration, further experiments are in progress. The results will be later used in isotherm, thermodynamic and kinetic calculations. There is also a need for more strict tests considering a diverse range of pollutants, different applied stresses and environmental conditions that can affect the separation performance of such an adsorbent. In this regard, software-assisted design of experiments could be helpful. It can reduce the consumption and waste of chemicals, help with regards to the eco-friendliness of chemical processes and actually shorten the pathway to industrial applications [27,28].

As an extra bonus, the enzyme immobilization successfully performed here can be promising for further applications of the buckypaper with respect to biosensing, e.g., the buckypaper adsorbent can potentially act as a biosensor as well. Adsorption of urease can change the electrical conductivity of the nanofilaments, and thus, the entire buckypaper. To validate this proof of concept, electrical conductivity of the buckypaper before and after adsorption of urease was recorded. As shown in Figure 5, CA functionalization can lower the electrical conductivity of the buckypaper due to the inclusion of carboxyl groups that act as electron-withdrawing elements, raising electrical resistivity. In contrast, adsorption of urease enhances the electrical conductivity notably, but with a lower rate for CA-functionalized buckypapers. One reason for the enhancement of conductivity could be the formation of electron transfer bridges between the nanofilaments by urease molecules. This observation can be interpreted in another way, i.e., bridging between enzyme molecules (i.e., the biosensing element) by the carbon nanofilaments. This phenomenon, i.e., the direct electrical connection of redox enzymes and electrodes through carbon nanomaterials have been reported earlier. Patolski et al. [29] showed this behavior through the alignment of glucose oxidase enzymes on the SWCNTs' tips that were structured as an array on a conductive substrate. Exposure of the enzyme-immobilized buckypaper to the urea, often monitored in blood to track kidney diseases, can alter the electrical conductivity and be considered as the sensing mechanism for such an analyte. It is worthy to note that the immobilization of enzymes is indeed the simplest technique that can tackle the bottleneck of their high solubility [30]. Enzyme immobilization allows for the tailoring of the bioreactions' conditions, and thus enables a continuous process with minimum pollution by the reaction products, an extremely desirable characteristic in the food industry. Moreover, it guarantees an improved stability, lifespan and ease of removal of the enzyme from the reaction medium at the end of the process, enabling cost effectiveness and recycling of the enzyme. As mentioned earlier, immobilization can also lead to stabilization of biocatalysts, prevent their unfolding and immunize the polypeptide bonds against rupture.

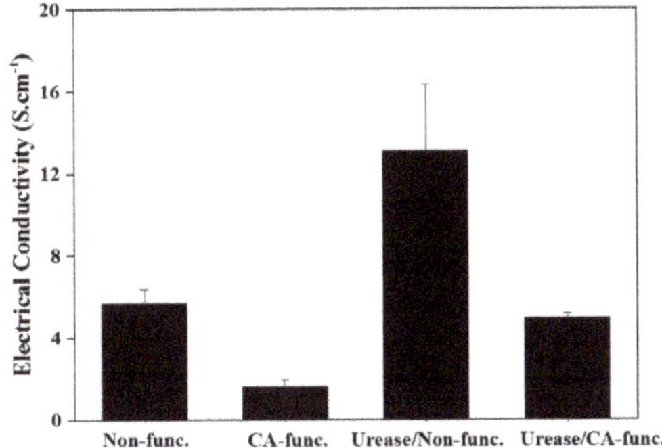

Figure 5. Electrical conductivity of the various classes of buckypapers before and after the adsorption of urease measured via a four-probe test.

4. Conclusion

Taken together, we devised a buckypaper adsorbent based on amphiphilic carbon nanofilaments that could separate urease molecules from water effectively (as large as 88%). The separation tests were performed under dynamic conditions that could challenge the adsorbent more strictly. Desirable selectivity and permeance (over 4 kL·h^{-1}m^{-2}bar^{-1}) of this novel adsorbent/membrane holds great promise for further development of the system for practical applications. Furthermore, firm immobilization of urease on conductive nanofilaments can assure the efficiency of a potential biosensing system. This proof of concept makes us optimistic with respect to the high potential of such nanomaterials for water treatment and biosensing in an industrial platform. However, first, we need to tackle some relevant bottlenecks for upscaling of their production. Electrospinning has shown to be a reliable method for large scale production of nanofibers, but post treatment (i.e., carbonization) of nanofibers must be performed in a controlled manner following a precise protocol that can govern a desirable chemistry for nanofibers. This step must be optimized and designed in a more economical way. For instance, the as-developed carbon nanofibers need to be stronger to exclude the chopping step, while maintaining their uniformity, porosity and more importantly functionality. We are at the beginning of the development of this system for water treatment and biosensing, but the obtained results encourage and motivate us to start further working on our material either as is or coupled with extra reactive agents.

Author Contributions: S.H. conceived the idea, prepared samples, performed characterizations and drafted the manuscript. M.E. was involved in development of the idea and analysis of the results.

Funding: M.E. appreciates the financial support provided through Aalto University, Academy of Finland, and Helmholtz Association (Grant No. VH-NG-523).

Acknowledgments: The authors would like to acknowledge Kristian Bühr for the design of the water permeance measurement set-up, and Joachim Koll for the bubble point test.

Conflicts of Interest: The authors declare no conflict of interest. The founding sponsors had no role in the design of the study; in the collection, analyses, or interpretation of data; in the writing of the manuscript, and in the decision to publish the results.

References

1. Ramakrishna, S.; Fujihara, K.; Teo, W.-E.; Yong, T.; Ma, Z.; Ramaseshan, R. Electrospun nanofibers: Solving global issues. *Mater. Today* **2006**, *9*, 40–50. [CrossRef]

2. Homaeigohar, S.; Dai, T.; Elbahri, M. Biofunctionalized nanofibrous membranes as super separators of protein and enzyme from water. *J. Colloid Interface Sci.* **2013**, *406*, 86–93. [CrossRef]

3. Homaeigohar, S.; Davoudpour, Y.; Habibi, Y.; Elbahri, M. The electrospun ceramic hollow nanofibers. *Nanomaterials* **2017**, *7*, 383. [CrossRef]

4. Homaeigohar, S.; Disci-Zayed, D.; Dai, T.; Elbahri, M. Biofunctionalized nanofibrous membranes mimicking carnivorous plants. *Bioinspir. Biomim. Nanobiomater.* **2013**, *2*, 186–193. [CrossRef]

5. Homaeigohar, S.; Elbahri, M. Nanocomposite electrospun nanofiber membranes for environmental remediation. *Materials* **2014**, *7*, 1017–1045. [CrossRef]

6. Homaeigohar, S.; Zillohu, A.U.; Abdelaziz, R.; Hedayati, M.K.; Elbahri, M. A novel nanohybrid nanofibrous adsorbent for water purification from dye pollutants. *Materials* **2016**, *9*, 848. [CrossRef]

7. Razali, M.; Kim, J.F.; Attfield, M.; Budd, P.M.; Drioli, E.; Lee, Y.M.; Szekely, G. Sustainable wastewater treatment and recycling in membrane manufacturing. *Green Chem.* **2015**, *17*, 5196–5205. [CrossRef]

8. Ma, Z.W.; Kotaki, M.; Ramakrishna, S. Electrospun cellulose nanofiber as affinity membrane. *J. Membr. Sci.* **2005**, *265*, 115–123. [CrossRef]

9. Bamford, C.H.; Allamee, K.G.; Purbrick, M.D.; Wear, T.J. Studies of a novel membrane for affinity separations: I. Functionalization and protein coupling. *J. Chromatogr.* **1992**, *606*, 19–31. [CrossRef]

10. Ma, Z.; Kotaki, M.; Ramakrishna, S. Surface modified nonwoven polysulphone (PSU) fiber mesh by electrospinning: A novel affinity membrane. *J. Membr. Sci.* **2006**, *272*, 179–187. [CrossRef]

11. Zhang, H.; Nie, H.; Yu, D.; Wu, C.; Zhang, Y.; White, C.J.B.; Zhu, L. Surface modification of electrospun polyacrylonitrile nanofiber towards developing an affinity membrane for bromelain adsorption. *Desalination* **2010**, *256*, 141–147. [CrossRef]

12. Elbahri, M.; Homaeigohar, S.; Dai, T.; Abdelaziz, R.; Khalil, R.; Zillohu, A.U. Smart metal-polymer bionanocomposites as omnidirectional plasmonic black absorbers formed by nanofluid filtration. *Adv. Funct. Mater.* **2012**, *22*, 4771–4777. [CrossRef]

13. Ma, Z.; Lan, Z.; Matsuura, T.; Ramakrishna, S. Electrospun polyethersulfone affinity membrane: Membrane preparation and performance evaluation. *J. Chromatogr. B-Anal. Technol. Biomed. Life Sci.* **2009**, *877*, 3686–3694. [CrossRef]

14. Lu, P.; Hsieh, Y.-L. Lipase bound cellulose nanofibrous membrane via Cibacron Blue F3GA affinity ligand. *J. Membr. Sci.* **2009**, *330*, 288–296. [CrossRef]

15. Homaeigohar, S.; Strunskus, T.; Strobel, J.; Kienle, L.; Elbahri, M. A flexible oxygenated carbographite nanofilamentous buckypaper as an amphiphilic membrane. *Adv. Mater. Interfaces* **2018**, *5*, 1800001. [CrossRef]

16. Homaeigohar, S.S.; Mahdavi, H.; Elbahri, M. Extraordinarily water permeable sol gel formed nanocomposite nanofibrous membranes. *J. Colloid Interface Sci.* **2012**, *366*, 51–56. [CrossRef]

17. Lee, H.D.; Kim, H.W.; Cho, Y.H.; Park, H.B. experimental evidence of rapid water transport through carbon nanotubes embedded in polymeric desalination membranes. *Small* **2014**, *10*, 2653–2660. [CrossRef]

18. Baker, R. *Membrane Technology and Applications*, 2nd ed.; Wiley: Chichester, UK, 2004.

19. Homaeigohar, S.S.; Buhr, K.; Ebert, K. Polyethersulfone electrsopun nanofibrous composite membrane for liquid filtration. *J. Membr. Sci.* **2010**, *365*, 68–77. [CrossRef]

20. Gao, H.; Kong, Y.; Cui, D.; Ozkan, C.S. Spontaneous insertion of DNA oligonucleotides into carbon nanotubes. *Nano Lett.* **2003**, *3*, 471–473. [CrossRef]

21. Pogorilyi, R.; Melnyk, I.; Zub, Y.; Seisenbaeva, G.; Kessler, V. Immobilization of urease on magnetic nanoparticles coated by polysiloxane layers bearing thiol-or thiol-and alkyl-functions. *J. Mater. Chem. B* **2014**, *2*, 2694–2702. [CrossRef]

22. Azevedo, A.M.; Prazeres, D.M.; Cabral, J.M.; Fonseca, L.P. Stability of free and immobilised peroxidase in aqueous–organic solvents mixtures. *J. Mol. Catal. B Enzym.* **2001**, *15*, 147–153. [CrossRef]

23. Villeneuve, P.; Muderhwa, J.M.; Graille, J.; Haas, M.J. Customizing lipases for biocatalysis: A survey of chemical, physical and molecular biological approaches. *J. Mol. Catal. B Enzym.* **2000**, *9*, 113–148. [CrossRef]

24. Hu, M.; Mi, B. Enabling graphene oxide nanosheets as water separation membranes. *Environ. Sci. Technol.* **2013**, *47*, 3715–3723. [CrossRef]

25. Singh, G.; Rana, D.; Matsuura, T.; Ramakrishna, S.; Narbaitz, R.M.; Tabe, S. Removal of disinfection byproducts from water by carbonized electrospun nanofibrous membranes. *Sep. Purif. Technol.* **2010**, *74*, 202–212. [CrossRef]

26. Barisci, J.N.; Tahhan, M.; Wallace, G.G.; Badaire, S.; Vaugien, T.; Maugey, M.; Poulin, P. Properties of carbon nanotube fibers spun from DNA-stabilized dispersions. *Adv. Funct. Mater.* **2004**, *14*, 133–138. [CrossRef]

27. Weissman, S.A.; Anderson, N.G. Design of experiments (DoE) and process optimization. A review of recent publications. *Org. Process Res. Dev.* **2015**, *19*, 1605–1633. [CrossRef]

28. Valtcheva, I.B.; Marchetti, P.; Livingston, A.G. Crosslinked polybenzimidazole membranes for organic solvent nanofiltration (OSN): Analysis of crosslinking reaction mechanism and effects of reaction parameters. *J. Membr. Sci.* **2015**, *493*, 568–579. [CrossRef]

29. Patolsky, F.; Weizmann, Y.; Willner, I. Long-range electrical contacting of redox enzymes by SWCNT connectors. *Angew. Chem.* **2004**, *116*, 2165–2169. [CrossRef]

30. Schoemaker, H.E.; Mink, D.; Wubbolts, M.G. Dispelling the myths—Biocatalysis in industrial synthesis. *Science* **2003**, *299*, 1694–1697. [CrossRef]

Review

An Overview of the Water Remediation Potential of Nanomaterials and Their Ecotoxicological Impacts

Mehrnoosh Ghadimi [1], **Sasan Zangenehtabar** [1] **and Shahin Homaeigohar** [2],*

[1] Department of Physical Geography, Faculty of Geography, University of Tehran, Tehran 1417466191, Iran; ghadimi@ut.ac.ir (M.G.); zangenegeo154@gmail.com (S.Z.)

[2] Nanochemistry and Nanoengineering, Department of Chemistry and Materials Science, School of Chemical Engineering, Aalto University, Kemistintie 1, 00076 Aalto, Finland

* Correspondence: Shahin.homaeigohar@aalto.fi

Received: 17 March 2020; Accepted: 15 April 2020; Published: 17 April 2020

Abstract: Nanomaterials, i.e., those materials which have at least one dimension in the 1–100 nm size range, have produced a new generation of technologies for water purification. This includes nanosized adsorbents, nanomembranes, photocatalysts, etc. On the other hand, their uncontrolled release can potentially endanger biota in various environmental domains such as soil and water systems. In this review, we point out the opportunities created by the use of nanomaterials for water remediation and also the adverse effects of such small potential pollutants on the environment. While there is still a large need to further identify the potential hazards of nanomaterials through extensive lab or even field studies, an overview on the current knowledge about the pros and cons of such systems should be helpful for their better implementation.

Keywords: nanomaterials; water treatment; environmental risks

---

1. Introduction

Water, a previously plentiful, free resource across the world, has become a rare, costly object over recent decades and currently, water shortage is going to be a challenge for sustainable development of the human community [1,2]. This crisis is dramatically expanding and is regarded a global systemic risk, mainly resulting from urban, agricultural, and industrial pollution. In these areas, water consumption has incremented up to 70% (agriculture), 22% (industry), and 8% (domestic) of the currently available fresh water and, accordingly, an enormous volume of wastewater containing a variety of pollutants has been produced [3]. No doubt, the release of wastewater from commercial and industrial sectors besides untreated domestic sewage and chemical pollutants into fresh water resources is horribly detrimental to both the human community and the ecosystem, including animals and plants. In this regard, the major water contaminants are heavy metal ions, organics (e.g., dyes), and oils that can disqualify any water stream for drinking.

To address the need for water remediation systems, during the past few decades, with the evolution of nanotechnology, a diverse range of new technologies based on nanomaterials has been developed. For instance, as adsorbent systems, nanomaterials offer an extremely large reactive surface area at a low mass, can be produced at a much less cost compared to activated carbon and they can remove pollutants efficiently [2]. In this regard, a plethora of nano-adsorbents in various forms and dimensionalities (D) like nanoparticles (0D), nanofibers and nanotubes (1D), nanosheets (2D), and nanoflowers (3D) has been investigated [2]. In terms of composition, the diversity is indeed extreme and many organic and inorganic nanomaterials have been synthesized that can help purify water streams. The separation mechanism can be based on chemical/physical affinity of the pollutant for the surface of the nanomaterial or through size exclusion of the pollutant by a porous nanomaterial system. In the latter case, nanomaterials act either as the main building block of the porous separator structure, as seen for nanofibrous microfiltration or ultrafiltration membranes, or as an additive to a polymeric thin film membrane to improve its hydrophilicity and thermomechanical properties. The previously mentioned separation processes such as adsorption or filtration only gather the pollutant molecules in solid form but never entirely "eliminate" or "decompose" them. This issue could be problematic because disposal of the obtained sludge and fouling of the filtration system is challenging [4]. For such reasons, the separation process should be complemented by degradation processes such as photocatalysis, sonocatalysis, and reductive degradation that allow the decomposition of organic pollutants into non-toxic metabolites. Other than environmental remediation, nanomaterials can also be efficiently applied for environmental control and construction of sensors that can detect even trace amounts of water pollutants.

Despite all the beneficial potentials that nanomaterials offer for the sake of water remediation and control, their unwanted and uncontrolled release can harm the environment and health of human beings, animals and plants. While nanomaterials are produced in different dimensionalities and aspect ratios, nanoparticles, i.e., 0D nanomaterials, are indeed the most challenging ones in terms of environmental risks. Originating from anthropogenic and natural sources, nanoparticles have always been present in the environment. Nanoparticles suspended in air are normally regarded as ultrafine particles, while the ones existing in soil and water form colloids [5]. In urban areas, particularly metropolitan areas, the emissions of vehicles fueled with diesel and gasoline as well as stationary combustion sources, generate a large amount of particulate materials of different sizes. In this context, the amount of the manufactured nanoparticles exceeds 36% of the entire particulate number concentrations [6]. Apart from synthetic nanoparticles, naturally formed nanoparticles are also present in the atmosphere, though in a much lower concentration compared to the manufactured ones [5]. In aquatic media, the term "colloid" is generally ascribed to those particles whose size varies in the range of 1 nm to 1 μm. Aquatic colloids consist of macromolecular organics, including peptides and proteins, humic and fulvic acids, and also colloidal inorganic materials composed of hydrous iron and manganese oxides. The small size, extensive surface area, high surface energy, quantum confinement, and conformational behavior of the colloids enable them to bind to various organic and inorganic contaminants [5]. Lastly, with respect to soils, there is a variety of natural nanoparticles comprising organic matter, clays, iron oxides, and other minerals that are decisive in diverse bio-geochemical processes. The soil colloids encompassing nanoparticles and their impact on soil development (pedogenesis) and soil structure (dispersion and crusting) have been a demanding research topic for decades. In this regard, synthetic nanomaterials and the soil colloids made thereof could drive and ease the transfer process of contaminants in soils [5].

In this review, we introduce potential applications of nanomaterials for water remediation, and on the other hand, discuss the possible routes of their release into different environmental sectors like soils and water bodies and their harmful effects. Search in the "ISI Web of Knowledge" database on the main topics to be discussed in the current review, i.e., "Nanomaterials for Water Treatment" and "Environmental Impacts of Nanomaterials", provided 222 and 63 relevant articles, respectively, for the 10 year time period starting in 2010 (Figure 1). As seen in this graph, the number of studies on the former topic, that is nanomaterials for water treatment, has increased significantly over time during the past ten years. In contrast, the second topic, that is the environmental impacts of nanomaterials, has been less extensively investigated. This fact suggests the need to dedicate further time and cost investment for the purpose of uncovering the potential impacts and risks of nanomaterials on the environment that have been somewhat overlooked.

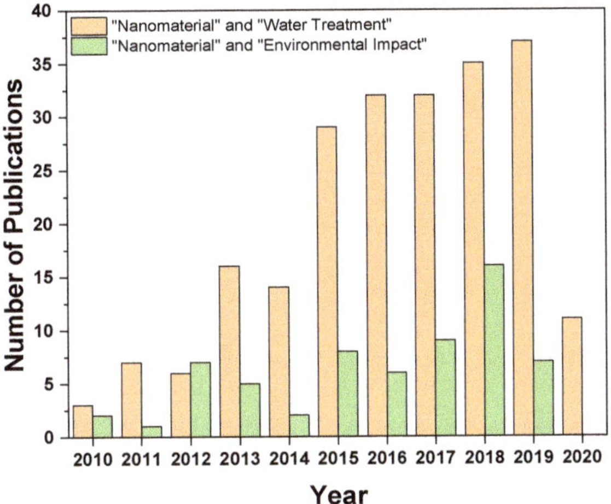

Figure 1. Annual number of publications in the last decade on the main keywords of the current review, according to ISI web of Knowledge (7 April 2020).

2. Nanomaterials for Water Purification

As mentioned earlier, nanomaterials offer several advantages for water treatment and control. This amazing potential stems from their large exposed surface area and functionality that can maximize their interactivity with water pollutants. In this section, we will take a glimpse on some important applications of nanomaterials at the service of water remediation.

2.1. Nanomaterials for Adsorption and Photodecomposition

In the water treatment field, the removal of dye pollutants due to their acute toxicities and carcinogenic nature is of paramount importance. Dyes have a history of thousands of years of application for textiles, paints, pigments, etc. Currently, almost 100,000 types of dyes are produced commercially. In terms of consumption volume, approximately 1.6 million tons of dyes are consumed annually. Thereof, 10–15% are wasted during use [7]. The dye pollutants released from industrial and agricultural wastes are refractory and potentially present carcinogenic effects. Therefore, they must be excluded from water streams through different kinds of traditional treatments, such as activated sludge, chemical coagulation, adsorption, and photocatalytic degradation [8]. Superior to the mentioned techniques, adsorption is relatively effective in the creation of a high quality effluent with no harmful byproducts in an energy/cost efficient manner [9–11]. This approach allows for exclusion of soluble and insoluble organic, inorganic, and biological water contaminants. The diverse merits of adsorption for dye removal are convincing enough to devise sustainable adsorbents that are manufactured on a large scale at low cost and enable fast and efficient dye removal. For this sake, within the course of the past few decades and with the evolution of nanotechnology, a variety of adsorbents of nanoscale size have been scrutinized. Nanomaterials provide an extensive reactive surface area at a low mass and versus activated carbon, i.e., the golden benchmark of adsorbents, they can be produced in a less expensive manner while removing dyestuffs and organic pollutants with a notably less amount [2]. Some examples of dye nano-adsorbents are as follow. Chitosan-coated magnetite (Fe_3O_4) nanoparticles showed a large adsorption capacity for crocein orange G (1883 mg/g) and acid green 25 (1471 mg/g). Interestingly, the adsorbent nanoparticles could be readily recovered by a magnetic field [12]. Based on such a concept, Fe_3O_4/activated carbon nanoparticles (6–16 nm) that can separate 138 and 166.6 mg/g methylene blue (MB) and brilliant green dyes, respectively, have been developed [13]. Dhananasekaran et al. [14] synthesized α-chitin nanoparticles (<50 nm) from Penaeus monodon shell waste and tested their dye (methylene blue (MB), bromophenol blue (BPB), and Coomassie brilliant blue (CBB)) adsorption efficiency. The nanoparticle adsorbent showed an adsorption efficiency of 95.96–99%, depending on the adsorbent concentration and based on physical adsorption of the dyestuff to the nanoparticles. Other than nanoparticles, nanofibrous adsorbents have also found application in the removal of dye pollutants from water. Such adsorbents are typically made through electrospinning and are superior to nanoparticulate adsorbents due to their easy recovery. As an example for nanofibrous adsorbents, polyethersulfone (PES) electrospun nanofibers containing V_2O_5 nanoparticles have been employed for removal of MB dye pollutant from water [15]. The nanocomposite nanofibers show a low isoelectric point thus at elevated pHs they acquire an extensive highly hydroxylated surface area that facilitates adsorption of cationic MB molecules.

One drawback of adsorption is its inability to completely "eliminate" or "decompose" the dye pollutants. Instead, it solely collects the dye molecules by transferring them to other phases. This feature could be challenging due to the fact that discharge of the dye-related sludge is not straightforward and the adsorbent is rarely reusable [4]. Accordingly, there is a need to complementary degrading treatments such as photocatalysis, sonocatalysis, and reductive degradation that enable decomposition of dye to non-toxic metabolites. In this regard, a variety of advanced oxidation processes (AOPs), provoking release of hydroxyl radicals (OH•), have shown a promising potential for decolorization of textile effluents. With unpaired electrons, OH• is drastically reactive and oxidizes recalcitrant organic pollutants [16]. Due to the abundance of low cost, while operative photocatalysts, photocatalysis is indeed of the most researched AOP processes and is considered as a practical degradation process for organic dyes and pesticides. Various semiconductor metal oxide nanoparticles such as ZnO and TiO_2 have shown notable efficiency in the photodecomposition of dye pollutants. For instance, Li et al. [17] developed an oil-in-water Pickering emulsion (PE) stabilized by the presence of TiO_2 particles, wherein the dye containing wastewater and insoluble organic matter were regarded as the water and oil phases, respectively. The TiO_2 particles could offer a large photoactivity effect and notably degrade the dye molecules (Figure 2a). In another relevant study, Kheirabadi et al. [18] synthesized a ternary nanostructure composed of Ag nanoparticle/ZnO nanorod/3-dimensional graphene (3DG) network via a coupled hydrothermal-photodeposition technique. While the 3DG can capture 300 mg/g MB dye by an adsorption process, Ag/ZnO component brings about the possibility of photodecomposition of the dye even under visible light irradiation. The dye removal mechanism of the synthesized adsorbent/photocatalytic system is illustrated in Figure 2b.

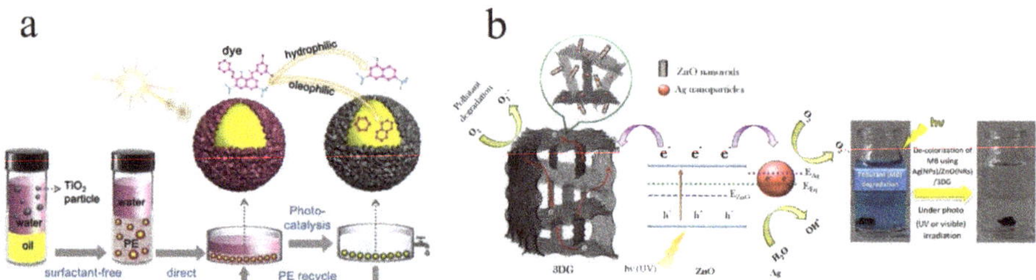

Figure 2. (a) Schematic illustration of an oil/water Pickering emulsion (PE) consisting superhydrophilic TiO_2 particles enabling dye photodecomposition. Reproduced with permission from [17]. Copyright 2019, Elsevier. (b) The dye removal mechanism of an adsorbent/photocatalyst system comprising Ag nanoparticle/ZnO nanorod/3D-graphene hydrogel. Reproduced with permission from [18]. Copyright 2019, Elsevier.

Despite the high potential of semiconductor materials for photodecomposition of various organic pollutants, commercial visible light photocatalysts suffer from poor stability or inefficiency upon irradiation. Addressing such challenges, group II–VI semiconductors whose energy gaps cover the visible light spectral range have been proposed as superior, compatible alternatives [3]. Large aggregation tendency, challenging separation and recovery are the other bottlenecks that have hampered the broader application of photocatalytic nanoparticles on a large, industrial scale [19,20]. One promising approach to overcome the abovementioned cons could be the nanocomposite strategy. By hybridizing the photocatalytic nanoparticles with polymeric nanofibers, not only is the large availability of the nanoparticles to the neighbouring water medium preserved, but also their intensive agglomeration is hampered and their recovery is facilitated. Nevertheless, due to the different polarity of the photocatalytic nanoparticles and polymeric nanofibers, the hybridization is not straightforward and can lead to clustering of the inorganic nanofillers in the polymer host [21]. In this regard, sol-gel treatment has shown promise in formation of tiny, isolated nanoparticles throughout the polymer nanofiber [22]. Even so, another important concern arises when recalling the possibility of photodecomposition of the encompassing polymer layer induced by the presence of photocatalytic nanoparticles [23]. Figure 3a–d show SEM images of the surface of TiO_2/PVC composite films after UV irradiation for different durations. According to these images, photodegradation of the PVC matrix is initiated from the PVC–TiO_2 interface and results in creation of cavities around TiO_2 particle aggregates that grow and in fact coalesce over time during the irradiation [24]. This behavior is also seen in nanocomposite nanofiber systems. For instance, TiO_2/PES nanofibers undergo photodegradation after exposure to UV irradiation, as reflected in their thermal and mechanical properties (Figure 3e,f, respectively [23]). Induced by UV irradiation, the electron/hole pairs formed in the conduction band (CB) and valence band (VB), respectively, react with O_2 and thereby create various active oxygen species such as O_2^-, $^1O_2$, $.O_2H$, and $.OH$ [24]. In the next step, these active oxygen species start the degradation process and attack the neighbouring polymer chains in the surface and later in the polymer bulk and deeper regions. When the carbon-centered radicals diffuse into the polymer chain, their successive reactions end up with the chain scission with the oxygen incorporation and CO_2 release [24].

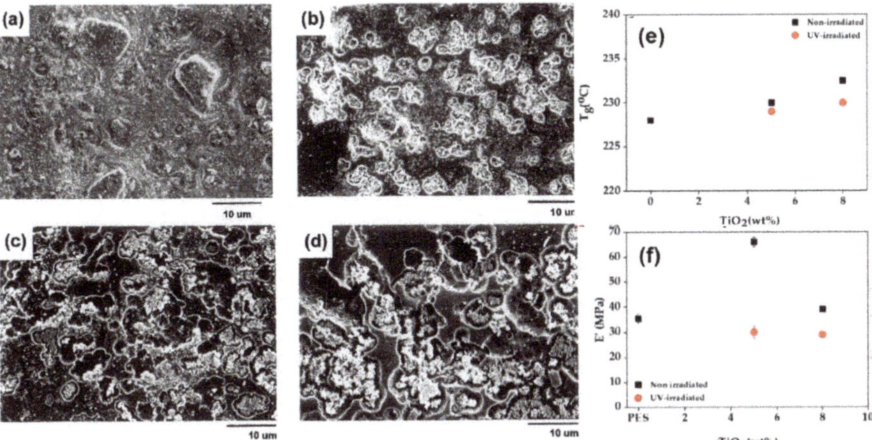

Figure 3. SEM images show the surface morphology of the TiO_2/PVC (1.5 wt.%) composite films after different irradiation times of (**a**) 0 h; (**b**) 25 h; (**c**) 50 h; (**d**) 100 h. Reproduced with permission from [24]. Copyright 2001, Elsevier. (**e**) Less notable increment of the glass transition temperature for the UV-irradiated TiO_2/PES nanocomposite nanofibers versus the non-irradiated ones. (**f**) The dynamic thermomechanical (DMTA) analysis implies that storage modulus for the nanocomposite nanofibers drops upon UV-irradiation. Reproduced with permission from [23]. Copyright the authors 2019, assigned to MDPI under a Creative Commons Attribution (CC BY) license.

2.2. Nanomaterials for Membrane-Based Water Treatment

A membrane is a selective barrier located between two homogenous phases that splits a feed water stream into a retentate and a permeate fraction. The pressure difference between the feed and permeate sides acts as the driving force for the membrane's action and passes water through the membrane [25]. As a result, based on charge, size, and shape, solutes and particles are discriminated (Figure 4a). The new generation of membrane technologies employ nanomaterials for water treatment.

Nanocomposite membranes comprising a thin polymeric film surface decorated or incorporated with nanofillers are a distinguished class of membranes able to dynamically purify water. Composite materials possess a favorable package of properties that are derived from a combination of encompassing medium's and filler's properties [26–36]. These properties are not restricted to the classic and predefined ones, rather new properties and functionalities arise, especially, when the filler's dimensions lie in the nanoscale. Nanomaterials in different forms and dimensionalities can be used in construction of nanocomposite membranes. Nanoparticles, for instance, have been widely used as nanofillers for mechanical reinforcement or for hydrophilization of polymeric membranes. In this regard, Rodrigues et al. [37] incorporated clay nanoparticles into mixed matrix polysulfone ultrafiltration membranes to improve thermomechanical properties and water permeability of the membrane, while maintaining optimum rejection efficiency. Moreover, the membranes reinforced with clay nanoparticles showed a lower fouling tendency and higher flux recovery when tested with sodium alginate and natural water. One critical concern regarding ultrafiltration (UF) membranes is their biofouling and the presence of bacterial colonies on the surface and thereby clogging the pores and lowering the permeability of the membrane. In this regard, *extracellular polymeric substances* (EPS) are released upon bacterial cell lysis and are adsorbed on the UF membrane and thus reduce the longevity and permeability of the membrane [38,39]. The most promising solution to address the challenge of biofouling of the UF membranes is surface hydrophilization by incorporation of various antifouling agents [40]. In this regard, a diverse range of antifouling agents has been employed in membrane technology, including Ag, Au, Cu, graphene oxide (GO), Zn, and TiO_2 nanoparticles [22,23], and also carbon nanotubes (CNTs) [38]. Despite the significance of industrial production of such nanocomposite membranes for water treatment, their toxicity that could originate from the release of the incorporated nanomaterials during the high pressure difference-driven filtration process should be carefully evaluated to minimize their adverse effects on human health and the environment [41]. The toxicity profile of the nanoparticles embedded in a polymeric matrix could be a function of their size, shape, charge and preparation conditions [38]. Among the nanofillers above mentioned, CNTs are resilient antibacterial agents whose toxic effect is derived from the ions and reactive oxygen species (ROSs) they release and thereby kill bacteria through oxidative stress stimuli [42]. Such a remarkable performance has led to wide application of CNTs in blended UF membranes, for the sake of improvement of filtration performance [43]. As reported in many studies, CNTs optimize water filtration and rejection of salts, nonpolar contaminants, micro- and macro-sized contaminants, and also waste chemical materials [44]. Ayyaru et al. [38] synthesized CNT- and sulphonated CNT (SCNT)-blended UF polyvinylidene fluoride (PVDF) membranes. For the latter group of the membranes, bovine serum albumin (BSA) rejection was 90%. As shown in Figure 4b, flux decline was less notable for the SCNT-PVDF membrane while permeating BSA solution through the membranes, thanks to its improved hydrophilicity. For the CNT- and SCNT-PVDF membranes, the fouling recovery ratio (FRR) was 72.74 and 83.52%, respectively, Figure 4c, implying their optimum antifouling effect arisen from $-SO_3H$ and $-OH$ groups found in SCNT and CNT, respectively. According to Figure 4d, the irreversible fouling value of the nanocomposite membranes, particularly that of SCNT-PVDF, is lower than that of the neat PVDF membrane. This again emphasizes the role of hydrophilicity induced by the presence of the nanofillers on lowering the fouling tendency of the membranes. Although CNTs are potentially versatile additives to membranes and also promising adsorbents for divalent metal ions, dyes, natural organic matters, etc., their relatively high unit cost is a limiting factor for their widespread practical use [3]. Moreover, the existence of metal catalysts in raw CNTs might induce a toxic effect. In contrast, chemically functionalized CNTs have not been shown yet to be toxic [45]. Accordingly, practical applicability of CNTs as adsorbents or inclusions in membranes for water treatment is tightly associated with finding cost effective production methods for CNTs and minimizing their toxicity effect by development of safer alternatives such as carbon nanocrystals (CNCs) [3].

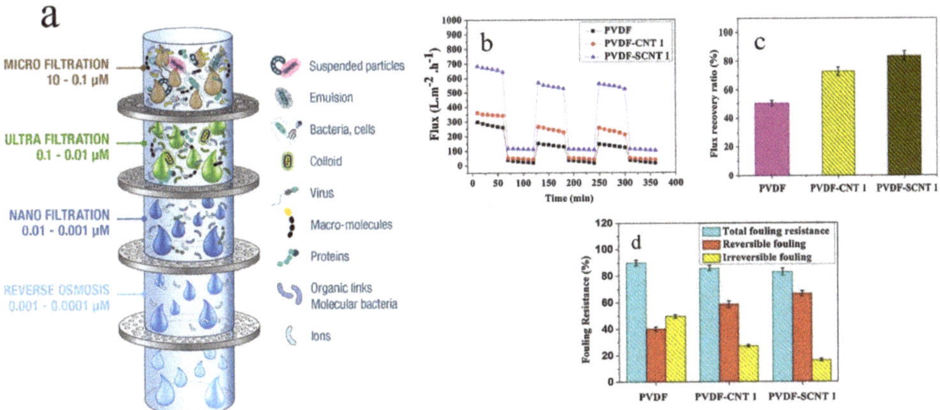

Figure 4. (**a**) Schematic shows diverse membrane filtration processes including reverse osmosis, nanofiltration, ultrafiltration, microfiltration, and particle filtration. Other than the reverse osmosis membranes whose structure is almost dense and non-porous, the rest are different in terms of the average pore size (the image was obtained from www.muchmorewater.com). Antifouling properties of CNT/PVDF UF membranes represented in cycle filtration (**b**), water recovery flux (**c**), and fouling resistance (**d**). Reproduced with permission from [38]. Copyright 2019, Elsevier.

Other than film-like membranes widely used for ultrafiltration and nanofiltration, electrospun nanofibrous membranes have also been developed for size exclusion of contaminants. Such nanostructured membranes possess a large porosity and an interconnected porous structure with microscale pore sizes. While the high porosity assures notable permeability to gas and liquid streams, the interconnected pores enhance fouling resistance. Such features lead to low energy consumption for the membrane process. Moreover, extensive available surface area and flexibility in surface functionality optimize the adsorptive nature and selectivity of the nanofibrous membranes. The constituting nanofibers could be as neat, chemically functionalized, nanocomposite, and even biofunctionalized [46–50]. For instance, a PES electrospun nanofiber mat overlaid on a poly(ethylene terephthalate) (PET) non-woven was evaluated as a membrane for liquid filtration and removal of micro- and submicron sized polystyrene (PS) particles from water [51]. Despite a high initial flux, upon increase of the feed pressure, the nanofibrous membrane's porosity declines and thereby the water flux drops. Therefore, while the nanofibrous membrane shows a high potential for pre-treatment of water e.g., as a microfiltration (MF) membrane, it should be mechanically reinforced and hydrophilized to raise water permeability and flux. To meet such objectives, PES nanofibers were stabilized and hydrophilized by incorporation of ZrO_2 [52] and TiO_2 [22] nanoparticles. Nanofibrous membranes can also be used for UF, i.e., a particular liquid filtration process separating a variety of pollutants, such as viruses, emulsions, proteins and colloids that are as small as 1–100 nm. For this sake, a nanofibrous membrane needs to possess a surface pore size less than 0.1 μm that alongside a high surface area potentially render them prone to rapid fouling and notable flux decline. To address this concern, a nanofibrous membrane is coated with a thin film and makes up a thin film composite (TFC) membrane [53]. Such a concept has shown applicability for forward osmosis (FO) membranes, as well [54].

Self-sustained hydrophilic nanofiber supports have been investigated for construction of the TFC FO membranes [55]. With a particular scaffold-like structure, the nanofiber support optimally lowers the internal concentration polarization (ICP) and raises water flux. To address the challenge of biofouling, the nanofibers could be equipped with antimicrobial properties, as well. For this sake, antibacterial nanoparticles e.g., Ag nanoparticles can be incorporated within the nanofibers. This strategy has been previously applied for various applications with respect to wound dressings [56] and water filtration [57]. For FO water treatment, Pan et al. [55] synthesized a TFC FO membrane based on an antibiofouling Ag nanoparticle-incorporated nanofibrous support layer, that could offer an improved water flux and reduce biofouling and ICP, Figure 5a–e. The as-formed FO membrane provides a remarkable bactericidal effect for *E. coli* (96%) and *S. aureus* (92%), thanks to release of Ag$^+$-species into the solution.

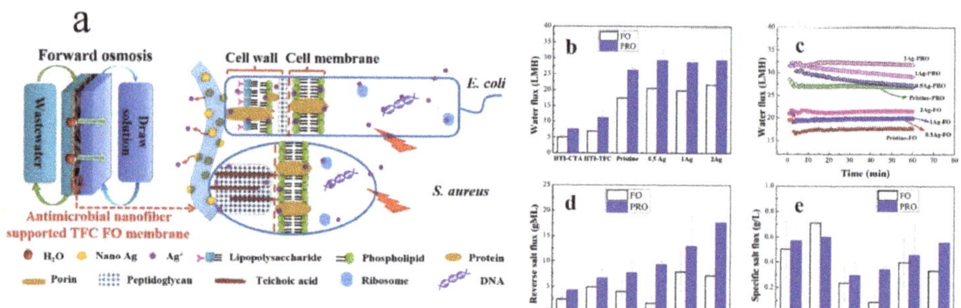

Figure 5. (**a**) Schematic illustration of the structure of the antimicrobial nanofiber supported TFC FO membrane (Ag/PAN-thin film nanocomposite (TFN)) (left) and the antibacterial action of the Ag nanoparticles incorporated in the nanofibers that damage DNA and membranes of the bacteria (right). (**b**) Water flux of the Ag/PAN-TFN versus that of commercial FO membranes and (**c**) water flux trend of the Ag/PAN-TFN FO membranes over time (in the FO and PRO modes). (**d**) Reverse salt flux and (**e**) specific salt flux of the Ag/PAN-TFN membrane compared to those of commercial FO membranes (in the FO and PRO modes). Note that the experiments were performed using 0.5 M NaCl as draw solution and DI water as feed solution; In the FO and PRO mode, the active layer face feed and draw solution, respectively. Reproduced with permission from [55]. Copyright 2019, American Chemical Society.

As shown in Figure 5a, the Ag ions are able to (quasi)covalently bond with thiols, phosphates, and organic amines available in proteins, lipopolysaccharide, and phospholipid of the cell membrane, and cell wall, thereby damaging them. Moreover, some ions could pass through the cell wall and adversely influence ribosomal subunit proteins and enzymes [58] and disrupt DNA's structure, leading to cell death. Other than the bactericidal activity, the hydrophilic, porous nanofibrous support allows for a superior water flux compared to commercial FO membranes (HTI-CTA and HTI-TFC) in two modes of FO and pressure-retarded osmosis (PRO, Figure 5b). The water flux remains steady over time in the FO mode, whereas it declines in the PRO mod due to dilution of the draw solution, leading to a lower osmotic pressure difference (Figure 5c). Regardless of the operation mode, the nanocomposite membranes show an increased reverse salt flux versus the commercial FO membranes (Figure 5d). In contrast, as shown in Figure 5e, the nanocomposite membranes exhibit a much less specific reverse salt flux, implying their higher selectivity and efficiency for a FO process.

Graphene, i.e., a 2D, 1-atom-thick planar sheet of sp^2 bonded carbon atoms, possesses remarkable physical, mechanical, thermal and optical properties [59]. In relation to water remediation, graphene's atomic thickness can potentially guarantee a high fluid permeability (that is significantly larger than that of typical nanofiltration (NF) membranes) and therefore lower energy consumption and inexpensive operation. Moreover, the 2D nanochannels forming between stacked graphene sheets or the nanopores available in a single graphene layer enable size-selective transport and purification of water streams (Figure 6a) [60]. The graphene that has been employed for development of water desalination membranes can be in different forms, such as pristine graphene, graphene oxide (GO) and reduced GO (rGO). Also, structurally graphene membranes can be constructed either as single layers or as stacked, multilayer forms. Monolayer graphene membranes with intrinsic pores have been studied for NF purpose experimentally as well as theoretically via simulation. For instance, O'Hern et al. [61] mounted a monolayer of chemical vapor deposition (CVD) graphene onto a porous polycarbonate substrate and thereby fabricated a graphene composite membrane with an active filtration area of

$25\,\text{mm}^2$. The graphene monolayer contained intrinsic nanopores as small as 1–15 nm that could contribute to the size-selective passage of molecules such as KCl, tetramethylammonium chloride, Allura Red AC (496 Da dye) and tetramethylrhodamine dextran (70 kDa), through the membrane made thereof. While KCl and tetramethylammonium chloride permeated through the graphene membrane, the diffusion of tetramethylrhodamine dextran was hampered. Despite the feasibility of selective molecular transport through the monolayer graphene membrane, selectivity is not controllable thanks to arbitrary sizes and locations of the intrinsic pores. Conclusively, formation of graphene layers with a large number of nanopores with adjusted, near monodisperse sizes and chemistries is a sophisticated objective that must be targeted in the new generation of graphene membranes.

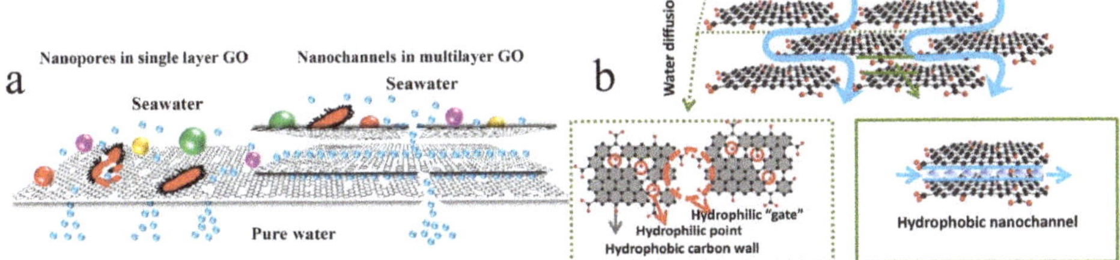

Figure 6. (**a**) Schematic illustration of the water purification mechanism for a single layer graphene membrane containing nanopores of tailored size and a multilayer graphene membrane comprising stacked GO sheets. Reproduced with permission [62]. Copyright 2018, Elsevier. (**b**) Schematic depicts the nanochannels forming between adjacent GO sheets that comprise hydrophobic and hydrophilic zones facilitating water flow and removal of tiny water pollutants. Reproduced with permission from [63]. Copyright 2016, American Chemical Society.

Despite various advantages of the monolayer graphene membranes for nanofiltration and even water desalination, large scale production of nanoporous graphene is indeed challenging and a single layer graphene is not robust enough to withstand the usual filtration pressures. In contrast, multi-layered GO membranes can be produced in a scalable manner and survive under large applied pressures. The nanochannels formed between the stacked GO nanosheets decorated with various polar functional groups allow water permeate through the membrane [64]. Induced by the notable slip length of water within the interlayer channels, the stacked GO inhibits passage of the solute particles. Figure 6b schematically shows how water molecules get into the hydrophilic zones and slip through the hydrophobic nanochannels. With respect to selectivity, because of hydration, the interlayer spacing of the GO nanosheets rises to 0.9 nm upon their immersion in ionic solutions. This structural change enables permeation of K^+ and Na^+ ions, and disqualifies the membrane for desalination purposes [65].

Despite the promising potential of graphene membranes for water purification, their release into water and thereby the environment is a concern that should be taken into account. In this regard, there are several reviews in the literature that widely discuss the fate, transformation, and toxicological impacts of such nanomaterials in the environment [66–68]. However, there is still a need to realistic, long term determination of the environmental implications of graphene nanomaterials. These precise ecotoxicological and life-cycle analyses enable us to better judge pros and cons of graphene nanomaterials and to find out how we can employ the safest ones with the least health and environmental concerns.

As highlighted so far, there are many promising achievements in relation to employment of nanomaterials for water treatment. Some of the recent developments (as of 2019) in this field are tabulated in Table 1. Many nanoadsorbents and nanomembranes are currently in development stage for the sake of large-scale production and industrialization. Given the fact that the water recycling is regarded an important aspect of sustainable development in the human communities, particularly when recalling the expanding water shortage crisis across the world, significance of realizing advanced water treatment technologies using nanomaterials is further stressed. However, this tiny functional building blocks could be also problematic and harmful in terms of sustainability, if released into the environment in an uncontrolled manner. In the next section, we review how they would be scattered in different media of water and soil and how they influence the biota living in such systems.

Table 1. Some examples of recent studies (as of 2019) on nanomaterials for water treatment.

Composition	Structure	Water Pollutant	Removal Mechanism	Nanomaterial Role	Ref
Cu NP/CNT/PVDF	Nanocomposite film	arsenic	Dynamic adsorption and oxidation	As oxidizer and adsorbent	[69]
Co doped ZrO_2	Nanoparticle	MO dye	Visible light photodegradation	As photocatalyst	[70]
NiO	Nanoparticle	ciprofloxacin	Adsorption	As adsorbent	[71]
Fe_3O_4 NP/AC	Nanocomposite particle	MO and RhB dye	Adsorption	To enable magnetic recovery and to raise adsorption capacity	[72]
Fe_3O_4@MIL-100(Fe)	Nanocomposite MOF	diclofenac sodium (DCF)	Adsorption and photodegradation	Magnetic recovery	[73]
$Fe_xCo_{3-x}O_4$	Nanoparticle	CR dye	Adsorption	To offer adsorption activity with easy magnetic recovery	[74]
ZnO-$ZnFe_2O_4$	Nanofiber	CR dye	Adsorption	To raise adsorption efficiency	[75]
Ag-ZnO/PANI	Nanocomposite film	BG dye	Adsorption	To raise adsorption efficiency	[76]
ZnS NP/PES	Film membrane	Humic acid	Filtration assisted by the antifoulant NPs	As antifouling agent	[77]
ZnO/KGM-PVA	Nanofiber membrane	MO dye	Visible light Photodegradation	To induce photocatalytic and antibacterial activity	[78]
Boehmite NP/EPVC	Nanocomposite Film membrane	BSA	Ultrafiltration	To improve hydrophilicity and water flux	[79]
Ag NP/wood	Nanocomposite Film membrane	MB dye	physical adsorption and catalytic degradation	Dye adsorption and antibacterial activity	[80]
(3-aminopropyl-triethoxysilane) APTES-Fe_3O_4 NP/PES	Nanocomposite Film membrane	arsenic	Adsorption	Heavy metal ion adsorption	[81]
PEI/PD/Ag NP	Nanocomposite Film membrane	BSA/HA/Oil	Ultrafiltration	As anti-fouling and anti-biofouling agent	[82]
Carbon dioxide plasma treated PVDF	Nanofiber membrane	CV dye and iron oxide NPs	size exclusion and adsorption	Ionic selectivity	[83]
Bentonite NP/PA	Nanocomposite Film membrane	NaCl	Reverse osmosis	To raise water permeability	[84]
PVA/PAN	Nanofiber membrane	Nanoparticles and Cr (VI) and Cd (II) ions	Adsorption and microfiltration	PVA nanofibers as the mechanical support and PAN nanofibers for selective adsorption of the ions	[85]
Clay NP/mixed matrix PS	Nanocomposite Film membrane	PEG and sodium alginate	Ultrafiltration	To improve antifouling properties, membrane thermal/ mechanical resistance and permeability with minimal loss in rejection	[37]
Clay NP/mixed matrix PS	Nanocomposite Film membrane	PEG and sodium alginate	Ultrafiltration	To improve antifouling properties, membrane thermal/ mechanical resistance and permeability with minimal loss in rejection	[37]
CS NP&Ag-CS NP/polyphenylsulfone	Nanocomposite Hollow fiber membrane	Reactive black dye	Adsorption	To improve porosity, dye rejection efficiency, hydrophilicity, and antifouling property	[86]

NP: nanoparticle; CNT: carbon nanotube; PVDF: polyvinylidene fluoride; Co: cobalt; MO: methyl orange; AC: activated carbon; RhB: Rhodamine B; MOF: metal-organic framework; CR: congo red; PANI: polyaniline; BG: brilliant green; AB: acid blue; AY: acid dye; PES: polyether sulfone; KGM: Konjac glucomannan; PVA: polyvinylalcohol; EPVC: emulsion polyvinyl chloride; BSA: bovine serum albumin; MB: methylene blue; PEI: poly(ether imide); PD: polydopamine; HA: humic acid; CV: crystal violet; PA: polyamide; PAN: polyacrylonitrile; PS: polysulfone; PEG: polyethylene glycol; CS: chitosan.

3. Ecotoxicology of Nanomaterials

As an inevitable result, with extensive use in water treatment, nanomaterials can be released into soils and water bodies, thereby threatening the quality of life of humans, animals, and plants (Figure 7a). In the case of uncontrolled release, they are accumulated and as suspended solids contaminate food and drinking water. Their final destiny strongly depends on their properties as well as the characteristics of the environment they are released into. Consequently, a variety of adverse ecotoxicological impacts can take place on microorganisms, plants, invertebrates, and fish. In the case of human, no significant risk from such nanomaterials has been reported, though the relevant studies performed so far are insufficient [87]. It is extremely necessary to research on the potentially hazardous effects of the nanomaterials on ecosystem and human health. This will encompass quantitative and qualitative assessment of such nanomaterials in different segments of the environment and determination of the likely consequences on the health of various species living in that segment. Accordingly, the research must be directed towards identification and testing of the environmental fate and transport, and ecotoxicology and toxicology of the nanomaterials.

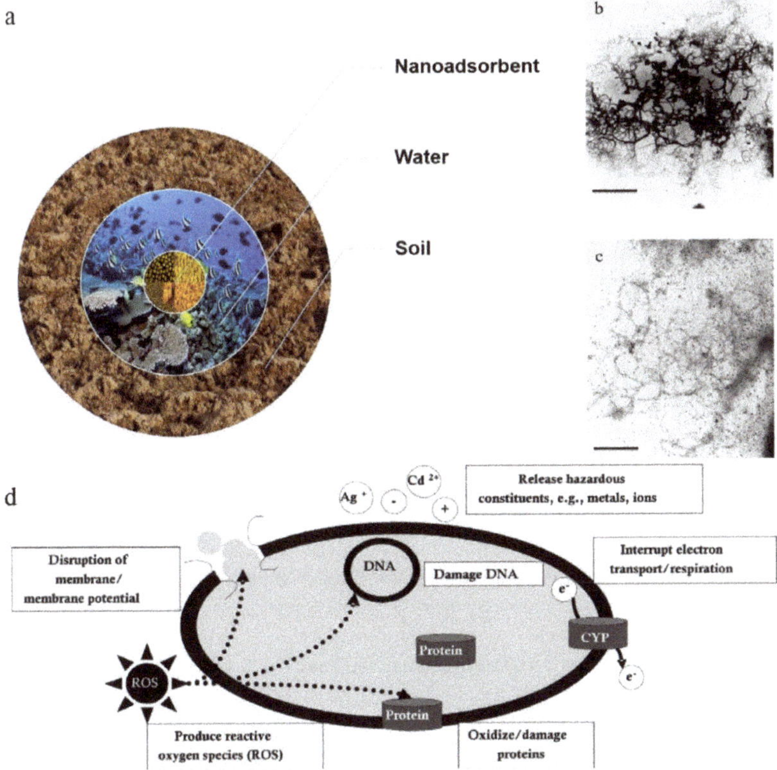

Figure 7. (**a**) The cycle of release of nanomaterials into environment from water to soil and vice versa. (**b**) The aquatic colloid comprising hydrous iron oxide aggregates and organic fibrils (scale bar = 1 µm). (**c**) The aquatic colloid made form silver stain on organic fibrils along with the polysaccharide fibrils possibly decorated with hydrous iron oxides (scale bar = 200 nm). (**d**) The likely cytotoxicity mechanisms of nanomaterials exerted on bacteria. CYP = cytochrome P. Reproduced with permission from [5]. Copyright 2009, John Wiley and Sons.

Among nanomaterials, carbon and metal oxide ones (e.g., TiO_2) are hardly biodegradable and persist in the environment [88,89]. It is anticipated that many of such insoluble nanomaterials would aggregate (as homo- or heteroaggregation) in the ecosystem and eventually settle out [90,91] and be subjected to various species as sediments. With respect to the soluble nanomaterials, the dissolution rate and solubility are determining in their fate. The release of toxic ions and generation of reactive oxygen species (ROS) are considered as the main toxicological pathways for metal/metal oxide nanomaterials, thereby threatening the life quality of aquatic and terrestrial creatures [5,92,93].

3.1. Nanomaterials in Aquatic Systems

As a general rule, the nanomaterials released into the environment interact with different substances available in that medium and thereby experience aggregation, dissolution, sedimentation, and transformation [94]. In 2010 it was reported that from 8 to 28% of the released nanomaterials are accommodated in soil, while between 4 and 7% in water [95]. These statistics are associated with the higher potential of soil (i.e., its constituents) and its interaction level with the liberated nanomaterials.

The nanomaterials are released into different ecosystems including the aquatic ones either intentionally or unintentionally. While the nanomaterial can be on purpose added to address the available contamination in the groundwater [96], they can also originate from the atmospheric emissions and solid/liquid waste streams delivered from factories, for instance. Furthermore, the nanoparticles present in paints, textile, and health care products, e.g., sunscreens and cosmetics, can potentially be released into the environment. As an example, thanks to desirable bactericidal ability, Ag nanoparticles are commonly employed in the consumer products (e.g., to inactivate the odor-causing bacteria in socks). However, their release into water streams can kill also useful bacteria that exclude ammonia from wastewater treatment systems [97]. Metal oxide nanoparticles, such as TiO_2, ZnO and CeO_2, are also extensively utilized in diverse products (e.g., sunscreens, paints, coatings, catalysts) and can reach water systems in different ways. The released nanoparticles are settled on soil and surface water systems, and in case of proper surface treatment they remain un-aggregated and float on the water. The nanoparticles that are deposited on the land, contaminate soil, transit through the surface, reach the ground water and in the course of this travel affect the existing biota. Wind and rainwater stream also displace the nanoparticles available in the solid wastes, wastewater effluents, direct or uncontrolled emissions into water bodies. Though accidental release of the nanomaterials can be strictly controlled within the production units, likely spillage during transportation to the consuming units is indeed challenging [5].

The majority of released nanoparticles aggregate as soon as hydrated, thereby being sediment in different rates. The extent of aggregation depends on the nanoparticle's surface charge and charge magnitude, and likely coverage of the nanoparticle's surface by mono- and divalent cations, by natural organic matter (NOM) or other organic molecules. In fact, such features determine the prevail of attractive and repulsive forces, thereby governing the nanoparticles' aggregation or their sticking to other surfaces [89] e.g., aquatic colloids.

Aquatic colloids, according to the definition established by the International Union of Pure and Applied Chemistry (IUPAC), are the materials whose dimension(s) are below 1 µm. Accordingly, aquatic colloids encompass (natural) nanoparticles, i.e., the materials with a minimum of two dimensions larger than 1 and smaller than 100 nm. In terms of composition, aquatic colloids are composed of organic (mainly humic substances and protein and polysaccharide fibrils) and inorganic fractions (e.g., metal oxides of Fe, Mn, and Al, and silicon oxide) and also microorganisms including viruses and bacteria (Figure 7b,c). It is very likely that nanoparticles (nanomaterials) interact with the aquatic colloids and thereby aggregate and sediment. This means that fate of the nanoparticle can be dominated by characteristics and concentration of aquatic colloids. For instance, it is reported that in estuarine and marine waters, density of aquatic colloids is notably low, on the other hand, concentration of nanoparticle is also low due to the high aggregation tendency and thus increased sedimentation rate at the aqueous systems with high ionic strengths [5]. It is worthy to note that the concentration of the nanoparticles commonly found in natural waters including those made of Ag and oxides of Ti, Ce, and Zn could range from 1 to 10 µg/L, and cumulatively, this concentration can reach 100 µg/L [5]. With respect to the interaction between the aquatic colloid and nanoparticles, humic substances coat the surface of the nanoparticle [98] and stabilize their surface charge, thereby minimizing the chance of aggregation [99], as shown for CNTs, for instance [100]. On the other hand, the fibrils raise this possibility via bridging mechanisms [101]. The main factors or properties of the nanoparticles that significantly influence their behavior in natural water systems are: chemical composition, mass, particle density, surface area, size distribution, surface charge, surface contamination (the likely shell and capping agents), and stability and solubility of the nanoparticle [5].

Given that industrial discharges are mostly exposed to marine environments, and the freshwater streams and coastal runoff end up to seas, this medium and its contamination is of utmost importance. The sea environment has a high ionic strength, is typically alkaline, and contains a diverse range of NOMs as well as colloids whose type and concentration depends on the location (coastal zone versus oceanic one). This medium can potentially be contaminated by the nanoparticles that are released via atmospheric deposition and/or coastal runoffs. The physicochemical properties of water such as temperature, salinity, and type of NOM varies by depth and affects the aggregation and colloid formation [102]. Similar to freshwater systems, the formed aggregates of the nanoparticulate contaminants precipitate slowly down to the ocean floor. In this route, they may stop at the interface of cold and warm streams or even be recycled by the existing biota. In either cases of complete sedimentation on the ocean floor or getting stuck in the mentioned interface, the living species corresponding to the zones might be affected. Additionally, the nanoparticles could be suspended and trapped in the surface microlayer of oceans and thereby impose risks to the birds, mammals and the species living in the microlayer [103].

The uptake of nanoparticles by living organisms can induce toxicity effects through different mechanisms. Nanoparticles can find a way into the cells by penetration through cell membranes, endocytosis as well as adhesion [104–106]. As soon as the nanoparticle is accommodated in the cell, it starts to damage the natural functions in different ways such as destruction of membrane structure or potential, proteins oxidation, genotoxicity, blockage of energy transduction, and generation of ROS and toxic substances [5]. Such mechanisms are illustrated schematically in Figure 7d. For instance, graphene is toxic to bacteria and damages the membrane and raises the oxidative stress level. In the graphene family, GO shows the most notable antibacterial activity, followed by rGO and graphite [107].

3.2. Nanomaterials in Terrestrial Systems

Given the large availability of a reactive sink, that can lead to overestimation of the exerted dose to the biota relative to the real one, soil is notably distinct from fresh and marine waters. In a similar manner, soil can be contaminated with nanomaterials intentionally (for the purpose of remediation, fertilizing, etc. e.g.) or unintentionally (by uncontrolled spills in the course of production and transport, e.g.) [5,108]. Regardless of the application aim, they can affect the biota present or dealing with soil.

Upon contacting soil, nanomaterials are physicochemically adsorbed to the soil particles' surface through hydrophobic interactions, hydrogen bonding, electrostatic interactions, etc. Induced by the presence of organics, they could experience chemical transformations. Additionally, nanomaterials can penetrate into the pores of macroparticles and stay there for a long time [109]. The behavior of nanomaterials i.e., their retention or mobility, in soil is dependent on several factors including soil texture, pH, humic acid, and chemistry of soil, surface coating and nanomaterial size, and pore water velocity. Particularly, pH and humic acid of soil determine the aggregation and colloidal stability of nanomaterials in soil [110].

Soil is a medium that accommodates diverse species such as microorganisms, plants and nematodes. These soil inhabitants are crucial in the cycle of nutrients, decomposition of materials, and nitrogen fixation [110]. Therefore, any damage to such important ecosystem elements can have non-compensable consequences on the life quality and nature.

Microorganisms are vital for performance and health of soil, taking into account their role in modulation of the organics' turnover and the cycle of mineral nutrients. They also play a crucial role in the physical characteristics of soil, thereby influencing water maintaining potential and the tendency of compaction or erosion. Accordingly, any undesired effect on microorganism community from the released nanomaterials could indirectly affect soil's quality and function. A diverse range of nanomaterials including fullerenes (C_{60} and nC_{60}) [111], 3-aminopropyl/silica, palladium, dodecanethiol/gold and copper nanoparticles [112] have been challenged with respect to their effect on soil's microbial community. In this regard, specifically, the factors such as soil respiration, microbial biomass, phospholipid fatty acid quantity, methyl ester of fatty acids, enzymatic activities, colony forming unit and DNA profile of bacterial community have been analyzed. While the above cited reports imply no toxicity of the mentioned nanomaterials, there are other studies such as that of Johansen et al. [113] on nC60 that explicitly demonstrate a notable decline in bacteria density of the soils exposed to this kind of nanomaterial. Nogueira et al. [108] also showed that the bacterial communities in soil samples treated with gold nanorods, TiO_2 nanoparticles, the polymeric nanomaterials composed of carboxylmethyl-cellulose (CMC), the hydrophobically modified CMC (HM-CMC), the hydrophobically modified polyethylglycol (HM-PEG), and the vesicles of sodium dodecyl sulphate/didodecyl dimethylammonium bromide (SDS/DDAB) are notably influenced by their toxicity effects. According to Rodrigues et al.'s study [114], bacteria can survive when exposed to SWCNTs, while fungal microbiota are unable to recover after the exposure. In another study, it was shown that FeO and Ag nanoparticles can decrease fungal biomass [115]. In contrast, quantum dots and super paramagnetic nanoparticles impose no notable toxicity to *Fusarium oxysporum* [116]. In general, the nitrifying bacteria are more vulnerable to nanomaterials' toxic effects compared to the bacteria with nitrogen fixing and denitrifying ability [110]. In the case of the ammonia oxidizing bacteria, several nanomaterials such as TiO_2 nanoparticles have shown a toxicity effect [117].

Plants are the other main species that are influenced by the presence of nanomaterials in the soil. Several kinds of nanomaterials have shown the ability to diffuse into plants via their roots, translocation, biotransformation, and spread in different forms across their structure, thereby impacting on photosynthesis, growth and regeneration abilities [118,119]. The resulting perturbation of the physiological functions impacts on seed germination, seedling growth, higher ROS production, damage to cell walls, and changes in proteins, carbohydrates, lipids, pigments, and hormones [120]. In this regard, metal oxide nanomaterials have been widely used as fertilizers for agriculture [121]. These nanoparticles account for a major number of the nanoparticle contaminants in soil including: Al_2O_3, TiO_2, CeO_2, ZnO, CuO and ZrO_2. The other important nanocontaminants are carbon fiber, SiO_2 and Ag nanoparticles, as well as carbon black [122]. Of the aforementioned metal oxide nanoparticles, ZnO and CeO_2 are especially detrimental for the plants [121]. As shown by Priester et al. [123] soybean (*Glycine max*) is vulnerable to high amounts of CeO_2 nanoparticles in soil, as witnessed by large ROS production, lipid peroxidation, visible stains (Figure 8a,b), and also declined total chlorophyll amounts. Compared to the leaves exposed to ZnO nanoparticles, the ones treated with CeO_2 nanoparticles show a higher percentage of visible damages, Figure 8c,d.

Feizi et al. [124] examined the effect of TiO_2 nanoparticles in different concentrations of 0, 5, 20, 40, 60 and 80 mg.L^{-1}, on the fennel seed germination. According to their results, after 2 weeks of seed incubation with TiO_2 nanoparticles (60 mg.L^{-1}), there was an increase in the total germination percentage. Other reports also show a rise in seed germination for tomato and rice when exposed to SiO_2 nanoparticles [125] and CNTs [126], respectively. It is worthy to note that concentration of the nanomaterial is an influential factor on the plant reaction and various plants might react differently to a single specific nanomaterial. It is imaginable that nanomaterial undergoes transformation within the plant and the resulting product can potentially impose a further risk or even be beneficial for the plant growth [110]. Other than the impacts on the growth process of plants, nanomaterials can affect and raise the generation of ROS in a plant. The high concentration of ROS can engender protein oxidation,

DNA and cell membrane destruction, lipid peroxidation, electrolyte leakage, etc., thereby inducing oxidative damage and eventually cell death [127].

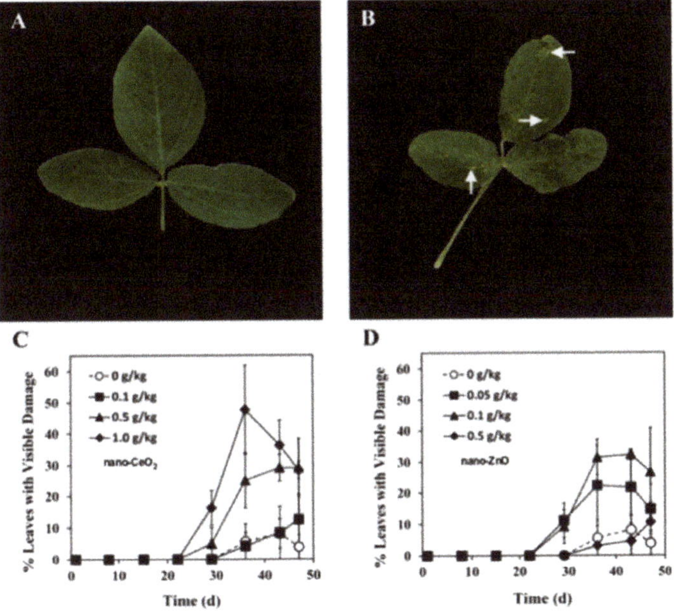

Figure 8. Camera images of soybean leaves without (**a**) and with (**b**) exposure to CeO_2 nanoparticles. The percentage of the leaves whose damage is recognized visually after exposure to (**c**) CeO_2 and (**d**) ZnO nanoparticles. Reproduced with permission from [123]. Copyright 2017, Elsevier.

One important point that must be kept in mind is that nanomaterials are normally transformed from origin i.e., the fabrication facility to destination. Thus, the results obtained in the studies based on pure nanomaterials could be not applicable to the realistic situations. Also, environmental factors such as pH, surface charge, ionic strength, surface coating, (UV) light irradiation, humic acids, inorganic ligands, mono- and divalent cations, affect the toxicity of nanomaterials in their destination [110]. With respect to the inclusion of nanoparticles into the terrestrial systems and their consequences, there are several reviews [110,128–131].

4. Conclusions and Future Outlook

Nanomaterials are favorable candidate materials for water remediation and control and an exciting prospect for their integration into point-of-use systems, and also in absolute removal of the current and emerging inorganic and organic pollutants from water is foreseen. They can be used in construction of nanoadsorbents that effectively capture polar and non-polar pollutants from water depending on their surface functionality. In this regard, the extensive surface area offered by nanomaterials is decisive and maximizes the adsorption efficiency. In case, the used nanomaterial is a photocatalyst, adsorption is extended to photodecomposition. Accordingly, the stuck organic pollutant is degraded into harmless byproducts and the adsorbent's surface becomes ready for a next round of adsorption/photodecomposition process. As another opportunity originated from nanomaterials, membrane nanostructures can be mentioned. Nanomaterials can be exploited as building blocks of a porous separator, as seen in electrospun nanofibrous membranes or single/few layer graphene membranes. Additionally, they can be used as additives to conventional thin film polymeric membranes for ultrafiltration to reduce fouling tendency and to raise thermomechanical properties. Despite such merits of nanomaterials in filtration and water treatment industry, there are a variety of bottlenecks that must be properly addressed:

(1) *Solar light driven photocatalysis*: With respect to photocatalysis-based water treatment, it is of paramount importance to avoid hole-electron recombination in the photocatalyst and also shift the light responsiveness from the UV to the visible solar light range. The latter goal guarrantees a lower energy consumption and wider applicability of photocatalysis for water purification. Research has begun to develop a new generation of solar light responsive doped photocatalysts that assure versatility and energy efficiency of photocatalytic nanoadsorbents.

(2) *Aggregation and poor recovery*: One important disadvantage regarding the nanoparticulate adsorbents is their aggregation tendency and challenging recovery. In this regard, one optimum solution is deposition of nanoparticles on nanostructured substrates e.g., nanofibers. This hybridization reduces the aggregation tendency and eases recovery of the nanoparticles, while preserving their high exposure to the external water medium.

(3) *Photodegradation of polymer hosts*: Many nanomaterials in different forms such as nanoparticles, nanotubes, nanofibers, and nanosheets are typically used as coupled with a polymer substrate or host. In case of applying photocatalytic, aggresive nanomaterials, the chance of photodegradation of the encapsulating polymer is considerable. To address this problem, inclusion of photostabilizers could be a main strategy.

(4) *Unwanted release of nanomaterials during the water treatment process*: Nanomaterials employed in the construction of micro-, ultra-, and nanofiltration membranes can be released into water streams when the membrane is subjected to harsh water streams and their complicated stress patterns. Therefore, primarily stabilization of nanoparticles on/in the membrane structure should be taken into account and secondly, intoxic materials should be employed that impose less hazardous effects on biota.

(5) *Long term, realistic testing of novel generation of nanostructured membranes*: Nanomaterials in higher dimensionalities such as 1D nanofibers and the 2D graphene family have also been studied for the development of membranes. Electrospun nanofibers have shown a promising potential in size exclusion and also adsorption of water pollutants. Thanks to their tunable pore size, high porosity and interconnected porous structure they can guarrantee a less energy consuming water treatment process. That is why they have found large applicability for building up advanced ultra- and nanofiltration membranes as a porous, robust support for the overlaying selective layer. However, no industrial utility has been reported for nanofibrous membranes. This could arises from the available gaps with respect to reliable testing of such membranes. Nanofibrous membranes must be challenged in long term, and under realistic conditions with real wastewater models and also be exposed to various complicated mechanical stress patterns. Typically, the relevant research experiments done at the lab scale consider only one type of pollutant and ignore co-existence of other dye, ionic, or organic pollutants, as seen in real wastewater, which compete for a limited number of available active/binding sites. Such a perspective was previously taken into account for activated carbon as a commercial adsorbent, and led to its commercialization. Graphene membranes are also a fascinating group of advanced nanomembranes that have shown amazing potentials, particlularly with respect to water permeability, while offering an ionic selectivity comparable to classic NF and ideally RO membranes. Nevertheless, their properties have been mainly theoretically validated rather than experimentally and there is still a large gap ahead till realistic employment of such membranes.

(6) *Large scale and economical production of nanostructured membranes and adsorbents*: This issue is under extensive investigation. In fact, technical difficulties with respect to scale-up and integration of nanomaterials into a relevant technology, cost effectiveness, and energy-related issue are all hindering concerns that have slowed the marketing trend of such products. For instance, TiO_2 nanoparticles and CNTs are among the most widely studied nanomaterials for adsorption of dyes. However, they are toxic and produced in a costly manner involving high temperature

and pressure. The former nano-adsorbent needs UV irradiation to photodecompose the dye pollutants that adds to the expenses of the treatment. In fact, it is highly necessary to produce large amounts of such nanomaterials at justifiable costs for water treatments, specific to different categories of wastewaters.

(7) *Environmental hazards*: This concern will persist in the future. This stems from the reality that many environmental and biological consequences of nanomaterials should be identified in the long term. Short term studies have shown that several nanomaterials are safe to human being, plants and animals. But, there is no certainty about their long term safety. For this reason, establishment of nanomaterial based water treatment systems should be followed with sufficient precautions. Technologically, it is also vital to secure such systems so that the release of nanomaterials into environment would be miminized.

Author Contributions: M.G. and S.Z. contributed to idea development and to drafting of the manuscript. S.H. conceived and developed the idea and drafted the manuscript. All authors have read and agreed to the published version of the manuscript.

Funding: This research received no external funding.

Acknowledgments: S.H. would like to acknowledge the financial support received from the European Union's Horizon 2020 research and innovation program under the Marie Sklodowska-Curie grant agreement No. 839165. Mady Elbahri is acknowledged for his support during the research.

Conflicts of Interest: The authors declare no conflict of interest.

References

1. Homaeigohar, S.; Elbahri, M. Nanocomposite Electrospun Nanofiber Membranes for Environmental Remediation. *Materials* **2014**, *7*, 1017–1045. [CrossRef] [PubMed]
2. Homaeigohar, S. The Nanosized Dye Adsorbents for Water Treatment. *Nanomaterials* **2020**, *10*, 295. [CrossRef] [PubMed]
3. Santhosh, C.; Velmurugan, V.; Jacob, G.; Jeong, S.K.; Grace, A.N.; Bhatnagar, A. Role of nanomaterials in water treatment applications: A review. *Chem. Eng. J.* **2016**, *306*, 1116–1137. [CrossRef]
4. Bansal, P.; Chaudhary, G.R.; Mehta, S.K. Comparative study of catalytic activity of ZrO2 nanoparticles for sonocatalytic and photocatalytic degradation of cationic and anionic dyes. *Chem. Eng. J.* **2015**, *280*, 475–485. [CrossRef]
5. Klaine, S.J.; Alvarez, P.J.J.; Batley, G.E.; Fernandes, T.F.; Handy, R.D.; Lyon, D.Y.; Mahendra, S.; McLaughlin, M.J.; Lead, J.R. Nanomaterials in the environment: Behavior, fate, bioavailability, and effects. *Environ. Toxicol. Chem.* **2008**, *27*, 1825–1851. [CrossRef]
6. Shi, J.P.; Evans, D.E.; Khan, A.; Harrison, R.M. Sources and concentration of nanoparticles (<10 nm diameter) in the urban atmosphere. *Atmos. Environ.* **2001**, *35*, 1193–1202.
7. Tan, K.B.; Vakili, M.; Horri, B.A.; Poh, P.E.; Abdullah, A.Z.; Salamatinia, B. Adsorption of dyes by nanomaterials: Recent developments and adsorption mechanisms. *Sep. Purif. Technol.* **2015**, *150*, 229–242. [CrossRef]
8. Shan, G.; Surampalli, R.Y.; Tyagi, R.D.; Zhang, T.C. Nanomaterials for environmental burden reduction, waste treatment, and nonpoint source pollution control: A review. *Front. Environ. Sci. Eng. China* **2009**, *3*, 249–264. [CrossRef]
9. Crini, G. Non-conventional low-cost adsorbents for dye removal: A review. *Bioresour. Technol.* **2006**, *97*, 1061–1085. [CrossRef]
10. Homaeigohar, S.; Elbahri, M. An Amphiphilic, Graphitic Buckypaper Capturing Enzyme Biomolecules from Water. *Water* **2019**, *11*, 2. [CrossRef]
11. Homaeigohar, S.; Strunskus, T.; Strobel, J.; Kienle, L.; Elbahri, M. A Flexible Oxygenated Carbographite Nanofilamentous Buckypaper as an Amphiphilic Membrane. *Adv. Mater. Interfaces* **2018**, *5*, 1800001. [CrossRef]
12. Chang, Y.C.; Chen, D.H. Adsorption kinetics and thermodynamics of acid dyes on a carboxymethylated chitosan-conjugated magnetic nano-adsorbent. *Macromol. Biosci.* **2005**, *5*, 254–261. [CrossRef] [PubMed]
13. Joshi, S.; Garg, V.K.; Kataria, N.; Kadirvelu, K. Applications of Fe3O4@AC nanoparticles for dye removal from simulated wastewater. *Chemosphere* **2019**, *236*, 124280. [CrossRef] [PubMed]

14. Dhananasekaran, S.; Palanivel, R.; Pappu, S. Adsorption of Methylene Blue, Bromophenol Blue, and Coomassie Brilliant Blue by α-chitin nanoparticles. *J. Adv. Res.* **2016**, *7*, 113–124. [CrossRef] [PubMed]

15. Homaeigohar, S.; Zillohu, A.U.; Abdelaziz, R.; Hedayati, M.K.; Elbahri, M. A Novel Nanohybrid Nanofibrous Adsorbent for Water Purification from Dye Pollutants. *Materials* **2016**, *9*, 848. [CrossRef] [PubMed]

16. Hisaindee, S.; Meetani, M.; Rauf, M. Application of LC-MS to the analysis of advanced oxidation process (AOP) degradation of dye products and reaction mechanisms. *Trac Trends Anal. Chem.* **2013**, *49*, 31–44. [CrossRef]

17. Li, Q.; Zhao, T.; Li, M.; Li, W.; Yang, B.; Qin, D.; Lv, K.; Wang, X.; Wu, L.; Wu, X.; et al. One-step construction of Pickering emulsion via commercial TiO2 nanoparticles for photocatalytic dye degradation. *Appl. Catal. B: Environ.* **2019**, *249*, 1–8. [CrossRef]

18. Kheirabadi, M.; Samadi, M.; Asadian, E.; Zhou, Y.; Dong, C.; Zhang, J.; Moshfegh, A.Z. Well-designed Ag/ZnO/3D graphene structure for dye removal: Adsorption, photocatalysis and physical separation capabilities. *J. Colloid Interface Sci.* **2019**, *537*, 66–78. [CrossRef]

19. Bhattacharyya, A.; Kawi, S.; Ray, M. Photocatalytic degradation of orange II by TiO2 catalysts supported on adsorbents. *Catal. Today* **2004**, *98*, 431–439. [CrossRef]

20. Gao, B.; Yap, P.S.; Lim, T.M.; Lim, T.-T. Adsorption-photocatalytic degradation of Acid Red 88 by supported TiO2: Effect of activated carbon support and aqueous anions. *Chem. Eng. J.* **2011**, *171*, 1098–1107. [CrossRef]

21. Meng, X.; Luo, N.; Cao, S.; Zhang, S.; Yang, M.; Hu, X. In-situ growth of titania nanoparticles in electrospun polymer nanofibers at low temperature. *Mater. Lett.* **2009**, *63*, 1401–1403. [CrossRef]

22. Homaeigohar, S.S.; Mahdavi, H.; Elbahri, M. Extraordinarily water permeable sol gel formed nanocomposite nanofibrous membranes. *J. Colloid Interface Sci.* **2012**, *366*, 51–56. [CrossRef] [PubMed]

23. Homaeigohar, S.; Botcha, N.K.; Zarie, E.S.; Elbahri, M. Ups and Downs of Water Photodecolorization by Nanocomposite Polymer Nanofibers. *Nanomaterials* **2019**, *9*, 250. [CrossRef] [PubMed]

24. Cho, S.; Choi, W. Solid-phase photocatalytic degradation of PVC–TiO2 polymer composites. *J. Photochem. Photobiol. A Chem.* **2001**, *143*, 221–228. [CrossRef]

25. Homaeigohar, S.; Elbahri, M. Graphene membranes for water desalination. *Npg Asia Mater.* **2017**, *9*, e427. [CrossRef]

26. Abadi, M.B.H.; Ghasemi, I.; Khavandi, A.; Shokrgozar, M.A.; Farokhi, M.; Homaeigohar, S.S.; Eslamifar, A. Synthesis of nano β-TCP and the effects on the mechanical and biological properties of β-TCP/HDPE/UHMWPE nanocomposites. *Polym. Compos.* **2010**, *31*, 1745–1753. [CrossRef]

27. Elbahri, M.; Zillohu, A.U.; Gothe, B.; Hedayati, M.K.; Abdelaziz, R.; El-Khozondar, H.J.; Bawa'aneh, M.; Abdelaziz, M.; Lavrinenko, A.; Zhukovsky, S.; et al. Photoswitchable molecular dipole antennas with tailored coherent coupling in glassy composite. *Light Sci. Appl.* **2015**, *4*, e316. [CrossRef]

28. Homaeigohar, S.S.; Sadi, A.Y.; Javadpour, J.; Khavandi, A. The effect of reinforcement volume fraction and particle size on the mechanical properties of β-tricalcium phosphate–high density polyethylene composites. *J. Eur. Ceram. Soc.* **2006**, *26*, 273–278. [CrossRef]

29. Homaeigohar, S.S.; Shokrgozar, M.A.; Sadi, A.Y.; Khavandi, A.; Javadpour, J.; Hosseinalipour, M. In vitro evaluation of biocompatibility of beta-tricalcium phosphate-reinforced high-density polyethylene; an orthopedic composite. *J. Biomed. Mater. Res. Part A* **2005**, *75A*, 14–22. [CrossRef]

30. Khalil, R.; Homaeigohar, S.; Häußler, D.; Elbahri, M. A shape tailored gold-conductive polymer nanocomposite as a transparent electrode with extraordinary insensitivity to volatile organic compounds (VOCs). *Sci. Rep.* **2016**, *6*, 33895. [CrossRef]

31. Homaeigohar, S.; Elbahri, M. Switchable Plasmonic Nanocomposites. *Adv. Opt. Mater.* **2019**, *7*, 1801101. [CrossRef]

32. Sadi, A.Y.; Homaeigohar, S.S.; Khavandi, A.R.; Javadpour, J. The effect of partially stabilized zirconia on the mechanical properties of the hydroxyapatite–polyethylene composites. *J. Mater. Sci. Mater. Med.* **2004**, *15*, 853–858. [CrossRef] [PubMed]

33. Sadi, A.Y.; Shokrgozar, M.; Homaeigohar, S.S.; Hosseinalipour, M.; Khavandi, A.; Javadpour, J. The effect of partially stabilized zirconia on the biological properties of HA/HDPE composites in vitro. *J. Mater. Sci. Mater. Med.* **2006**, *17*, 407–412. [CrossRef] [PubMed]

34. Yari Sadi, A.; Shokrgozar, M.A.; Homaeigohar, S.S.; Khavandi, A. Biological evaluation of partially stabilized zirconia added HA/HDPE composites with osteoblast and fibroblast cell lines. *J. Mater. Sci. Mater. Med.* **2008**, *19*, 2359–2365. [CrossRef]

35. Homaeigohar, S.S.; Shokrgozar, M.; Javadpour, J.; Khavandi, A.; Sadi, A.Y. Effect of reinforcement particle size on in vitro behavior of β-tricalcium phosphate-reinforced high-density polyethylene: A novel orthopedic composite. *J. Biomed. Mater. Res. Part A* **2006**, *78*, 129–138. [CrossRef]

36. Homaeigohar, S.; Kabir, R.; Elbahri, M. Size-tailored physicochemical properties of Monodisperse polystyrene nanoparticles and the nanocomposites Made thereof. *Sci. Rep.* **2020**, *10*, 1–11. [CrossRef]

37. Rodrigues, R.; Mierzwa, J.C.; Vecitis, C.D. Mixed matrix polysulfone/clay nanoparticles ultrafiltration membranes for water treatment. *J. Water Process Eng.* **2019**, *31*, 100788. [CrossRef]

38. Ayyaru, S.; Pandiyan, R.; Ahn, Y.-H. Fabrication and characterization of anti-fouling and non-toxic polyvinylidene fluoride -Sulphonated carbon nanotube ultrafiltration membranes for membrane bioreactors applications. *Chem. Eng. Res. Des.* **2019**, *142*, 176–188. [CrossRef]

39. Ayyaru, S.; Choi, J.; Ahn, Y.-H. Biofouling reduction in a MBR by the application of a lytic phage on a modified nanocomposite membrane. *Environ. Sci. Water Res. Technol.* **2018**, *4*, 1624–1638. [CrossRef]

40. Zodrow, K.; Brunet, L.; Mahendra, S.; Li, D.; Zhang, A.; Li, Q.; Alvarez, P.J. Polysulfone ultrafiltration membranes impregnated with silver nanoparticles show improved biofouling resistance and virus removal. *Water Res.* **2009**, *43*, 715–723. [CrossRef]

41. Yong, C.W. Study of interactions between polymer nanoparticles and cell membranes at atomistic levels. *Philos. Trans. R. Soc. B Biol. Sci.* **2015**, *370*, 20140036. [CrossRef] [PubMed]

42. Akhavan, O.; Ghaderi, E. Toxicity of graphene and graphene oxide nanowalls against bacteria. *Acs Nano* **2010**, *4*, 5731–5736. [CrossRef] [PubMed]

43. Al-Khaldi, F.A.; Abusharkh, B.; Khaled, M.; Atieh, M.A.; Nasser, M.; Saleh, T.A.; Agarwal, S.; Tyagi, I.; Gupta, V.K. Adsorptive removal of cadmium (II) ions from liquid phase using acid modified carbon-based adsorbents. *J. Mol. Liq.* **2015**, *204*, 255–263.

44. Das, R.; Ali, M.E.; Hamid, S.B.A.; Ramakrishna, S.; Chowdhury, Z.Z. Carbon nanotube membranes for water purification: A bright future in water desalination. *Desalination* **2014**, *336*, 97–109. [CrossRef]

45. Chen, C.; Wang, X. Adsorption of Ni (II) from aqueous solution using oxidized multiwall carbon nanotubes. *Ind. Eng. Chem. Res.* **2006**, *45*, 9144–9149. [CrossRef]

46. Homaeigohar, S.; Tsai, T.-Y.; Young, T.-H.; Yang, H.J.; Ji, Y.-R. An electroactive alginate hydrogel nanocomposite reinforced by functionalized graphite nanofilaments for neural tissue engineering. *Carbohydr. Polym.* **2019**, *224*, 115112. [CrossRef]

47. Homaeigohar, S.; Dai, T.; Elbahri, M. Biofunctionalized nanofibrous membranes as super separators of protein and enzyme from water. *J. Colloid Interface Sci.* **2013**, *406*, 86–93. [CrossRef]

48. Homaeigohar, S.; Disci-Zayed, D.; Dai, T.; Elbahri, M. Biofunctionalized nanofibrous membranes mimicking carnivorous plants. *Bioinspired Biomim. Nanobiomater.* **2013**, *2*, 186–193. [CrossRef]

49. Homaeigohar, S.; Davoudpour, Y.; Habibi, Y.; Elbahri, M. The Electrospun Ceramic Hollow Nanofibers. *Nanomaterials* **2017**, *7*, 383. [CrossRef]

50. Homaeigohar, S. Amphiphilic Oxygenated Amorphous Carbon-Graphite Buckypapers with Gas Sensitivity to Polar and Non-Polar VOCs. *Nanomaterials* **2019**, *9*, 1343. [CrossRef]

51. Homaeigohar, S.S.; Buhr, K.; Ebert, K. Polyethersulfone electrsopun nanofibrous composite membrane for liquid filtration. *J. Mem. Sci.* **2010**, *365*, 68–77. [CrossRef]

52. Homaeigohar, S.S.; Elbahri, M. Novel compaction resistant and ductile nanocomposite nanofibrous microfiltration membranes. *J. Colloid Interface Sci.* **2012**, *372*, 6–15. [CrossRef] [PubMed]

53. Yoon, K.; Hsiao, B.S.; Chu, B. Functional nanofibers for environmental applications. *J. Mater. Chem.* **2008**, *18*, 5326–5334. [CrossRef]

54. Song, X.; Liu, Z.; Sun, D.D. Energy recovery from concentrated seawater brine by thin-film nanofiber composite pressure retarded osmosis membranes with high power density. *Energy Environ. Sci.* **2013**, *6*, 1199–1210. [CrossRef]

55. Pan, S.-F.; Ke, X.-X.; Wang, T.-Y.; Liu, Q.; Zhong, L.-B.; Zheng, Y.-M. Synthesis of Silver Nanoparticles Embedded Electrospun PAN Nanofiber Thin-Film Composite Forward Osmosis Membrane to Enhance Performance and Antimicrobial Activity. *Ind. Eng. Chem. Res.* **2019**, *58*, 984–993. [CrossRef]

56. Homaeigohar, S.; Boccaccini, A.R. Antibacterial Biohybrid Nanofibers for Wound Dressings. *Acta Biomater.* **2020**, *107*, 25–49. [CrossRef]

57. Liu, L.; Liu, Z.; Bai, H.; Sun, D.D. Concurrent filtration and solar photocatalytic disinfection/degradation using high-performance Ag/TiO2 nanofiber membrane. *Water Res.* **2012**, *46*, 1101–1112. [CrossRef]

58. Yamanaka, M.; Hara, K.; Kudo, J. Bactericidal actions of a silver ion solution on Escherichia coli, studied by energy-filtering transmission electron microscopy and proteomic analysis. *Appl. Environ. Microbiol.* **2005**, *71*, 7589–7593. [CrossRef]

59. Liu, W.-W.; Chai, S.-P.; Mohamed, A.R.; Hashim, U. Synthesis and characterization of graphene and carbon nanotubes: A review on the past and recent developments. *J. Ind. Eng. Chem.* **2014**, *20*, 1171–1185. [CrossRef]

60. Han, Y.; Xu, Z.; Gao, C. Ultrathin graphene nanofiltration membrane for water purification. *Adv. Funct. Mater.* **2013**, *23*, 3693–3700. [CrossRef]

61. O'Hern, S.C.; Stewart, C.A.; Boutilier, M.S.; Idrobo, J.-C.; Bhaviripudi, S.; Das, S.K.; Kong, J.; Laoui, T.; Atieh, M.; Karnik, R. Selective molecular transport through intrinsic defects in a single layer of CVD graphene. *Acs Nano* **2012**, *6*, 10130–10138. [CrossRef] [PubMed]

62. Anand, A.; Unnikrishnan, B.; Mao, J.-Y.; Lin, H.-J.; Huang, C.-C. Graphene-based nanofiltration membranes for improving salt rejection, water flux and antifouling–A review. *Desalination* **2018**, *429*, 119–133. [CrossRef]

63. Wang, J.; Zhang, P.; Liang, B.; Liu, Y.; Xu, T.; Wang, L.; Cao, B.; Pan, K. Graphene oxide as an effective barrier on a porous nanofibrous membrane for water treatment. *ACS Appl. Mater. Interfaces* **2016**, *8*, 6211–6218. [CrossRef] [PubMed]

64. Dai, H.; Xu, Z.; Yang, X. Water permeation and ion rejection in layer-by-layer stacked graphene oxide nanochannels: A molecular dynamics simulation. *J. Phys. Chem. C* **2016**, *120*, 22585–22596. [CrossRef]

65. Joshi, R.; Carbone, P.; Wang, F.; Kravets, V.; Su, Y.; Grigorieva, I.; Wu, H.; Geim, A.; Nair, R. Precise and ultrafast molecular sieving through graphene oxide membranes. *Science* **2014**, *343*, 752–754. [CrossRef]

66. Hu, X.; Zhou, Q. Health and ecosystem risks of graphene. *Chem. Rev.* **2013**, *113*, 3815–3835. [CrossRef]

67. Bussy, C.; Ali-Boucetta, H.; Kostarelos, K. Safety considerations for graphene: Lessons learnt from carbon nanotubes. *Acc. Chem. Res.* **2012**, *46*, 692–701. [CrossRef]

68. Bianco, A. Graphene: Safe or toxic? The two faces of the medal. *Angew. Chem. Int. Ed.* **2013**, *52*, 4986–4997. [CrossRef]

69. Luan, H.; Teychene, B.; Huang, H. Efficient removal of As(III) by Cu nanoparticles intercalated in carbon nanotube membranes for drinking water treatment. *Chem. Eng. J.* **2019**, *355*, 341–350. [CrossRef]

70. Reddy, C.V.; Reddy, I.N.; Reddy, K.R.; Jaesool, S.; Yoo, K. Template-free synthesis of tetragonal Co-doped ZrO2 nanoparticles for applications in electrochemical energy storage and water treatment. *Electrochim. Acta* **2019**, *317*, 416–426. [CrossRef]

71. Rahdar, S.; Rahdar, A.; Igwegbe, C.A.; Moghaddam, F.; Ahmadi, S. Synthesis and physical characterization of nickel oxide nanoparticles and its application study in the removal of ciprofloxacin from contaminated water by adsorption: Equilibrium and kinetic studies. *Desalin. Water Treat.* **2019**, *141*, 386–393. [CrossRef]

72. Liu, X.; Tian, J.; Li, Y.; Sun, N.; Mi, S.; Xie, Y.; Chen, Z. Enhanced dyes adsorption from wastewater via Fe3O4 nanoparticles functionalized activated carbon. *J. Hazard. Mater.* **2019**, *373*, 397–407. [CrossRef] [PubMed]

73. Li, S.; Cui, J.; Wu, X.; Zhang, X.; Hu, Q.; Hou, X. Rapid in situ microwave synthesis of Fe3O4@MIL-100(Fe) for aqueous diclofenac sodium removal through integrated adsorption and photodegradation. *J. Hazard. Mater.* **2019**, *373*, 408–416. [CrossRef] [PubMed]

74. Liu, J.; Wang, N.; Zhang, H.; Baeyens, J. Adsorption of Congo red dye on FexCo3-xO4 nanoparticles. *J. Environ. Manag.* **2019**, *238*, 473–483. [CrossRef]

75. Appiah-Ntiamoah, R.; Baye, A.F.; Gadisa, B.T.; Abebe, M.W.; Kim, H. In-situ prepared ZnO-ZnFe2O4 with 1-D nanofiber network structure: An effective adsorbent for toxic dye effluent treatment. *J. Hazard. Mater.* **2019**, *373*, 459–467. [CrossRef] [PubMed]

76. Gouthaman, A.; Asir, J.A.; Gnanaprakasam, A.; Sivakumar, V.M.; Thirumarimurugan, M.; Ahamed, M.A.R.; Azarudeena, R.S. Enhanced dye removal using polymeric nanocomposite through incorporation of Ag doped ZnO nanoparticles: Synthesis and characterization. *J. Hazard. Mater.* **2019**, *373*, 493–503. [CrossRef]

77. Guo, J.; Khan, S.; Cho, S.-H.; Kim, J. ZnS nanoparticles as new additive for polyethersulfone membrane in humic acid filtration. *J. Ind. Eng. Chem.* **2019**, *79*, 71–78. [CrossRef]

78. Lv, D.; Wang, R.; Tang, G.; Mou, Z.; Lei, J.; Han, J.; De Smedt, S.; Xiong, R.; Huang, C. Ecofriendly electrospun membranes loaded with visible-light-responding nanoparticles for multifunctional usages: Highly efficient air filtration, dye scavenging, and bactericidal activity. *Acs Appl. Mater. Interfaces* **2019**, *11*, 12880–12889. [CrossRef]

79. Farjami, M.; Moghadassi, A.; Vatanpour, V.; Hosseini, S.M.; Parvizian, F. Preparation and characterization of a novel high-flux emulsion polyvinyl chloride (EPVC) ultrafiltration membrane incorporated with boehmite nanoparticles. *J. Ind. Eng. Chem.* **2019**, *72*, 144–156. [CrossRef]

80. Che, W.; Xiao, Z.; Wang, Z.; Li, J.; Wang, H.; Wang, Y.; Xie, Y. Wood-based mesoporous filter decorated with silver nanoparticles for water purification. *ACS Sustain. Chem. Eng.* **2019**, *7*, 5134–5141. [CrossRef]

81. Rowley, J.; Abu-Zahra, N.H. Synthesis and characterization of polyethersulfone membranes impregnated with (3-aminopropyltriethoxysilane) APTES-Fe3O4 nanoparticles for As(V) removal from water. *J. Environ. Chem. Eng.* **2019**, *7*, 102875. [CrossRef]

82. Saraswathi, M.S.S.A.; Rana, D.; Alwarappan, S.; Gowrishankar, S.; Vijayakumar, P.; Nagendran, A. Polydopamine layered poly (ether imide) ultrafiltration membranes tailored with silver nanoparticles designed for better permeability, selectivity and antifouling. *J. Ind. Eng. Chem.* **2019**, *76*, 141–149. [CrossRef]

83. Gopakumar, D.A.; Arumukhan, V.; Gelamo, R.V.; Pasquini, D.; de Morais, L.C.; Rizal, S.; Hermawan, D.; Nzihou, A.; Khalil, H.P.S.A. Carbon dioxide plasma treated PVDF electrospun membrane for the removal of crystal violet dyes and iron oxide nanoparticles from water. *Nano-Struct. Nano-Objects* **2019**, *18*, 100268. [CrossRef]

84. Kadhom, M.; Deng, B. Thin film nanocomposite membranes filled with bentonite nanoparticles for brackish water desalination: A novel water uptake concept. *Microporous Mesoporous Mater.* **2019**, *279*, 82–91. [CrossRef]

85. Liu, X.; Jiang, B.; Yin, X.; Ma, H.; Hsiao, B.S. Highly permeable nanofibrous composite microfiltration membranes for removal of nanoparticles and heavy metal ions. *Sep. Purif. Technol.* **2020**, *233*, 115976. [CrossRef]

86. Kolangare, I.M.; Isloor, A.M.; Karim, Z.A.; Kulal, A.; Ismail, A.F.; Asiri, A.M. Antibiofouling hollow-fiber membranes for dye rejection by embedding chitosan and silver-loaded chitosan nanoparticles. *Environ. Chem. Lett.* **2019**, *17*, 581–587. [CrossRef]

87. Boxall, A.B.; Tiede, K.; Chaudhry, Q. Engineered nanomaterials in soils and water: How do they behave and could they pose a risk to human health? *Nanomedicine* **2007**, *2*, 919–927. [CrossRef]

88. Flores-Cervantes, D.X.; Maes, H.M.; Schäffer, A.; Hollender, J.; Kohler, H.-P.E. Slow biotransformation of carbon nanotubes by horseradish peroxidase. *Environ. Sci. Technol.* **2014**, *48*, 4826–4834. [CrossRef]

89. Keller, A.A.; Wang, H.; Zhou, D.; Lenihan, H.S.; Cherr, G.; Cardinale, B.J.; Miller, R.; Ji, Z. Stability and aggregation of metal oxide nanoparticles in natural aqueous matrices. *Environ. Sci. Technol.* **2010**, *44*, 1962–1967. [CrossRef]

90. Adeleye, A.S.; Keller, A.A. Long-term colloidal stability and metal leaching of single wall carbon nanotubes: Effect of temperature and extracellular polymeric substances. *Water Res.* **2014**, *49*, 236–250. [CrossRef]

91. Conway, J.R.; Adeleye, A.S.; Gardea-Torresdey, J.; Keller, A.A. Aggregation, dissolution, and transformation of copper nanoparticles in natural waters. *Environ. Sci. Technol.* **2015**, *49*, 2749–2756. [CrossRef]

92. Wu, B.; Torres-Duarte, C.; Cole, B.J.; Cherr, G.N. Copper oxide and zinc oxide nanomaterials act as inhibitors of multidrug resistance transport in sea urchin embryos: Their role as chemosensitizers. *Environ. Sci. Technol.* **2015**, *49*, 5760–5770. [CrossRef] [PubMed]

93. Hong, J.; Rico, C.M.; Zhao, L.; Adeleye, A.S.; Keller, A.A.; Peralta-Videa, J.R.; Gardea-Torresdey, J.L. Toxic effects of copper-based nanoparticles or compounds to lettuce (Lactuca sativa) and alfalfa (Medicago sativa). *Environ. Sci. Process. Impacts* **2015**, *17*, 177–185. [CrossRef] [PubMed]

94. Markus, A.; Parsons, J.; Roex, E.; De Voogt, P.; Laane, R. Modeling aggregation and sedimentation of nanoparticles in the aquatic environment. *Sci. Total Environ.* **2015**, *506*, 323–329. [CrossRef] [PubMed]

95. Keller, A.A.; McFerran, S.; Lazareva, A.; Suh, S. Global life cycle releases of engineered nanomaterials. *J. Nanopart. Res.* **2013**, *15*, 1692. [CrossRef]

96. Zhang, W.X.; Elliott, D.W. Applications of iron nanoparticles for groundwater remediation. *Remediat. J.* **2006**, *16*, 7–21. [CrossRef]

97. Elbahri, M.; Homaeigohar, S.; Dai, T.; Abdelaziz, R.; Khalil, R.; Zillohu, A.U. Smart Metal-Polymer Bionanocomposites as Omnidirectional Plasmonic Black Absorbers Formed by Nanofluid Filtration. *Adv. Funct. Mater.* **2012**, *22*, 4771–4777. [CrossRef]

98. Gibson, C.T.; Turner, I.J.; Roberts, C.J.; Lead, J.R. Quantifying the Dimensions of Nanoscale Organic Surface Layers in Natural Waters. *Environ. Sci. Technol.* **2007**, *41*, 1339–1344. [CrossRef]

99. Tipping, E. The adsorption of aquatic humic substances by iron oxides. *Geochim. Cosmochim. Acta* **1981**, *45*, 191–199. [CrossRef]

100. Hyung, H.; Fortner, J.D.; Hughes, J.B.; Kim, J.-H. Natural organic matter stabilizes carbon nanotubes in the aqueous phase. *Environ. Sci. Technol.* **2007**, *41*, 179–184. [CrossRef]

101. Buffle, J.; Wilkinson, K.J.; Stoll, S.; Filella, M.; Zhang, J. A generalized description of aquatic colloidal interactions: The three-colloidal component approach. *Environ. Sci. Technol.* **1998**, *32*, 2887–2899. [CrossRef]

102. Yamashita, Y.; Tsukasaki, A.; Nishida, T.; Tanoue, E. Vertical and horizontal distribution of fluorescent dissolved organic matter in the Southern Ocean. *Mar. Chem.* **2007**, *106*, 498–509. [CrossRef]

103. Wurl, O.; Obbard, J.P. A review of pollutants in the sea-surface microlayer (SML): A unique habitat for marine organisms. *Mar. Pollut. Bull.* **2004**, *48*, 1016–1030. [CrossRef] [PubMed]

104. Lin, S.; Keskar, G.; Wu, Y.; Wang, X.; Mount, A.S.; Klaine, S.J.; Moore, J.M.; Rao, A.M.; Ke, P.C. Detection of phospholipid-carbon nanotube translocation using fluorescence energy transfer. *Appl. Phys. Lett.* **2006**, *89*, 143118. [CrossRef]

105. Kim, J.S.; Yoon, T.-J.; Yu, K.N.; Kim, B.G.; Park, S.J.; Kim, H.W.; Lee, K.H.; Park, S.B.; Lee, J.-K.; Cho, M.H. Toxicity and tissue distribution of magnetic nanoparticles in mice. *Toxicol. Sci.* **2005**, *89*, 338–347. [CrossRef] [PubMed]

106. Geiser, M.; Rothen-Rutishauser, B.; Kapp, N.; Schürch, S.; Kreyling, W.; Schulz, H.; Semmler, M.; Hof, V.I.; Heyder, J.; Gehr, P. Ultrafine particles cross cellular membranes by nonphagocytic mechanisms in lungs and in cultured cells. *Environ. Health Perspect.* **2005**, *113*, 1555–1560. [CrossRef]

107. Liu, S.; Zeng, T.H.; Hofmann, M.; Burcombe, E.; Wei, J.; Jiang, R.; Kong, J.; Chen, Y. Antibacterial activity of graphite, graphite oxide, graphene oxide, and reduced graphene oxide: Membrane and oxidative stress. *ACS Nano* **2011**, *5*, 6971–6980. [CrossRef]

108. Nogueira, V.; Lopes, I.; Rocha-Santos, T.; Santos, A.L.; Rasteiro, G.M.; Antunes, F.; Gonçalves, F.; Soares, A.M.V.M.; Cunha, A.; Almeida, A.; et al. Impact of organic and inorganic nanomaterials in the soil microbial community structure. *Sci. Total Environ.* **2012**, *424*, 344–350. [CrossRef]

109. Jaisi, D.P.; Elimelech, M. Single-walled carbon nanotubes exhibit limited transport in soil columns. *Environ. Sci. Technol.* **2009**, *43*, 9161–9166. [CrossRef]

110. Reddy, P.V.L.; Hernandez-Viezcas, J.A.; Peralta-Videa, J.R.; Gardea-Torresdey, J.L. Lessons learned: Are engineered nanomaterials toxic to terrestrial plants? *Sci. Total Environ.* **2016**, *568*, 470–479. [CrossRef]

111. Tong, Z.; Bischoff, M.; Nies, L.; Applegate, B.; Turco, R.F. Impact of Fullerene (C60) on a Soil Microbial Community. *Environ. Sci. Technol.* **2007**, *41*, 2985–2991. [CrossRef] [PubMed]

112. Shah, V.; Belozerova, I. Influence of metal nanoparticles on the soil microbial community and germination of lettuce seeds. *Water Air Soil Pollut.* **2009**, *197*, 143–148. [CrossRef]

113. Johansen, A.; Pedersen, A.L.; Jensen, K.A.; Karlson, U.; Hansen, B.M.; Scott-Fordsmand, J.J.; Winding, A. Effects of C60 fullerene nanoparticles on soil bacteria and protozoans. *Environ. Toxicol. Chem. Int. J.* **2008**, *27*, 1895–1903. [CrossRef]

114. Rodrigues, D.F.; Jaisi, D.P.; Elimelech, M. Toxicity of functionalized single-walled carbon nanotubes on soil microbial communities: Implications for nutrient cycling in soil. *Environ. Sci. Technol.* **2012**, *47*, 625–633. [CrossRef]

115. Feng, Y.; Cui, X.; He, S.; Dong, G.; Chen, M.; Wang, J.; Lin, X. The role of metal nanoparticles in influencing arbuscular mycorrhizal fungi effects on plant growth. *Environ. Sci. Technol.* **2013**, *47*, 9496–9504. [CrossRef] [PubMed]

116. Rispail, N.; De Matteis, L.; Santos, R.; Miguel, A.S.; Custardoy, L.; Testillano, P.S.; Risueño, M.C.; Pérez-de-Luque, A.; Maycock, C.; Fevereiro, P. Quantum dot and superparamagnetic nanoparticle interaction with pathogenic fungi: Internalization and toxicity profile. *Acs Appl. Mater. Interfaces* **2014**, *6*, 9100–9110. [CrossRef]

117. Luo, Z.; Qiu, Z.; Chen, Z.; Du Laing, G.; Liu, A.; Yan, C. Impact of TiO2 and ZnO nanoparticles at predicted environmentally relevant concentrations on ammonia-oxidizing bacteria cultures under ammonia oxidation. *Environ. Sci. Pollut. Res.* **2015**, *22*, 2891–2899. [CrossRef] [PubMed]

118. Ma, Y.; Zhang, P.; Zhang, Z.; He, X.; Li, Y.; Zhang, J.; Zheng, L.; Chu, S.; Yang, K.; Zhao, Y. Origin of the different phytotoxicity and biotransformation of cerium and lanthanum oxide nanoparticles in cucumber. *Nanotoxicology* **2015**, *9*, 262–270. [CrossRef]

119. Li, K.-E.; Chang, Z.-Y.; Shen, C.-X.; Yao, N. Toxicity of nanomaterials to plants. In *Nanotechnology and Plant Sciences*; Springer: Berlin/Heidelberg, Germany, 2015; pp. 101–123.

120. Majumdar, S.; Peralta-Videa, J.R.; Bandyopadhyay, S.; Castillo-Michel, H.; Hernandez-Viezcas, J.-A.; Sahi, S.; Gardea-Torresdey, J.L. Exposure of cerium oxide nanoparticles to kidney bean shows disturbance in the plant defense mechanisms. *J. Hazard. Mater.* **2014**, *278*, 279–287. [CrossRef]

121. Du, W.; Tan, W.; Peralta-Videa, J.R.; Gardea-Torresdey, J.L.; Ji, R.; Yin, Y.; Guo, H. Interaction of metal oxide nanoparticles with higher terrestrial plants: Physiological and biochemical aspects. *Plant Physiol. Biochem.* **2017**, *110*, 210–225. [CrossRef]

122. Holden, P.A.; Klaessig, F.; Turco, R.F.; Priester, J.H.; Rico, C.M.; Avila-Arias, H.; Mortimer, M.; Pacpaco, K.; Gardea-Torresdey, J.L. Evaluation of exposure concentrations used in assessing manufactured nanomaterial environmental hazards: Are they relevant? *Environ. Sci. Technol.* **2014**, *48*, 10541–10551. [CrossRef] [PubMed]

123. Priester, J.H.; Moritz, S.C.; Espinosa, K.; Ge, Y.; Wang, Y.; Nisbet, R.M.; Schimel, J.P.; Susana Goggi, A.; Gardea-Torresdey, J.L.; Holden, P.A. Damage assessment for soybean cultivated in soil with either CeO2 or ZnO manufactured nanomaterials. *Sci. Total Environ.* **2017**, *579*, 1756–1768. [CrossRef]

124. Feizi, H.; Rezvani Moghaddam, P.; Shahtahmassebi, N.; Fotovat, A. Impact of Bulk and Nanosized Titanium Dioxide (TiO2) on Wheat Seed Germination and Seedling Growth. *Biol. Trace Elem. Res.* **2012**, *146*, 101–106. [CrossRef] [PubMed]

125. Siddiqui, M.H.; Al-Whaibi, M.H. Role of nano-SiO2 in germination of tomato (Lycopersicum esculentum seeds Mill.). *Saudi J. Biol. Sci.* **2014**, *21*, 13–17. [CrossRef] [PubMed]

126. Nair, R.; Mohamed, M.S.; Gao, W.; Maekawa, T.; Yoshida, Y.; Ajayan, P.M.; Kumar, D.S. Effect of Carbon Nanomaterials on the Germination and Growth of Rice Plants. *J. Nanosci. Nanotechnol.* **2012**, *12*, 2212–2220. [CrossRef] [PubMed]

127. Sharma, P.; Jha, A.B.; Dubey, R.S.; Pessarakli, M. Reactive Oxygen Species, Oxidative Damage, and Antioxidative Defense Mechanism in Plants under Stressful Conditions. *J. Bot.* **2012**, *2012*, 26. [CrossRef]

128. Gardea-Torresdey, J.L.; Rico, C.M.; White, J.C. Trophic Transfer, Transformation, and Impact of Engineered Nanomaterials in Terrestrial Environments. *Environ. Sci. Technol.* **2014**, *48*, 2526–2540. [CrossRef]

129. Keller, A.A.; Fournier, E.; Fox, J. Minimizing impacts of land use change on ecosystem services using multi-criteria heuristic analysis. *J. Environ. Manag.* **2015**, *156*, 23–30. [CrossRef]

130. McKee, M.S.; Filser, J. Impacts of metal-based engineered nanomaterials on soil communities. *Environ. Sci. Nano* **2016**, *3*, 506–533. [CrossRef]

131. Kwak, J.I.; An, Y.-J. The current state of the art in research on engineered nanomaterials and terrestrial environments: Different-scale approaches. *Environ. Res.* **2016**, *151*, 368–382. [CrossRef]

Review

Role of Nanomaterials in the Treatment of Wastewater: A Review

Asim Ali Yaqoob [1], Tabassum Parveen [2], Khalid Umar [1,*] and Mohamad Nasir Mohamad Ibrahim [1,*]

[1] School of Chemical Sciences, Universiti Sains Malaysia, Penang 11800, Malaysia; asim.yaqoob@student.usm.my

[2] Department of Botany, Aligarh Muslim University, Aligarh 202002, India; tab_parveen@rediffmail.com

* Correspondence: khalidumar4@gmail.com (K.U.); mnm@usm.my (M.N.M.I.)

Received: 27 December 2019; Accepted: 8 February 2020; Published: 12 February 2020

Abstract: Water is an essential part of life and its availability is important for all living creatures. On the other side, the world is suffering from a major problem of drinking water. There are several gases, microorganisms and other toxins (chemicals and heavy metals) added into water during rain, flowing water, etc. which is responsible for water pollution. This review article describes various applications of nanomaterial in removing different types of impurities from polluted water. There are various kinds of nanomaterials, which carried huge potential to treat polluted water (containing metal toxin substance, different organic and inorganic impurities) very effectively due to their unique properties like greater surface area, able to work at low concentration, etc. The nanostructured catalytic membranes, nanosorbents and nanophotocatalyst based approaches to remove pollutants from wastewater are eco-friendly and efficient, but they require more energy, more investment in order to purify the wastewater. There are many challenges and issues of wastewater treatment. Some precautions are also required to keep away from ecological and health issues. New modern equipment for wastewater treatment should be flexible, low cost and efficient for the commercialization purpose.

Keywords: nanocatalysts; nanomembranes; nanosorbents; nanomaterial applications; waste water treatment; nanomaterial challenges

---

1. Introduction

Water is a natural source on the earth and its availability in pure state is very essential for human beings as well as for other living creatures because the concept of life is unbelievable without water. Water is also called universal solvent due to its potential properties like solubility power etc. Currently, major problem of whole world is water contamination, due to several reasons like inadequate sewage treatment, industrial wastes, marine dumping issues, radioactive waste material, some agricultural perspectives etc. [1,2]. Water pollution has an adverse effect on environment, and it can also responsible for air pollution that reflects very dangerous results on human health. Water pollution also carries an adverse impact on economic growth and socials perspectives of the concern societies/countries. Recently, a UN report stated that purified and freshwater availability is a global issue and become a challenge in 21st century for whole world because the survival of living creatures is not safe with contaminated water [3,4]. Contaminated water means that unwanted materials come into water bodies or reservoirs and make it unsuitable for drinking and other purposes. To overcome this emerging problem, there are many chemical, physical and mechanical methods. Moreover, researchers are still working by exploring different new technologies to improve water purification methods with low cost [5,6]. Newly, emerging field nanotechnology provides a potential offer to purify water with a low expense, high working efficiency in removing pollutants and reusable ability [7]. In past era, nanomaterials are successfully applied to several places like in a field of medical science, catalysis, etc.

Recently, when the world facing serious issues of drinking water, experts found that nanomaterial is better option to treat wastewater because it has some unique properties like nano size, large surface area, highly reactive, strong solution mobility [8], strong mechanical property, porosity characters, hydrophilicity, dispersibility and hydrophobicity [9–11]. Some heavy metal like Pb, Mo, etc. organic and inorganic pollutants and various harmful microbes are reported to be successfully removed by using different nanomaterials [12–16]. Currently, WHO (World Health Organization) reported that almost 1.7 million people died due to water pollution and four billion cases of different health issues were reported annually due to waterborne diseases [17]. Table 1 indicates different kinds of water pollutants with sources and their adverse effects.

Table 1. Indicates different water pollutants with their sources and adverse effects.

Water Pollutants	Source of Pollutants	Effect of Pollutants	References
Pathogens	Viruses and bacteria	Water borne diseases	[18]
Agricultural Pollutants	Agricultural chemicals	Directly affect the fresh water resources	[19]
Sediments and suspended solids	Land cultivation, demolition, mining operations	Damaging fish spawning, affecting aquatic environment of insects and fishes	[20]
Inorganic Pollutants	Metals compounds, Trace elements, Inorganic salts, Heavy metals, Mineral acids,	Aquatic flora and fauna, Public Health problems	[21]
Organic Pollutants	Detergents, Insecticides, Herbicides.	Aquatic life problems, Cacogenic.	[22]
Industrial Pollutants	Municipal pollutant Water	Caused water and air pollution.	[23]
Radioactive Pollutants	Different Isotopes	Bones, teeth, skin and can cause	[24]
Nutrients Pollutants	Plant debris, fertilizer.	Effect on eutrophication process.	[25]
Macroscopic Pollutants	Marine debris	Plastic pollution.	[26]
Sewage and contaminated water	Domestic Wastewater	Water borne diseases	[27]

Recently, there are more advance developments occurred in nanomaterials such as nanophotocatalysts, nanomotors, nanomembranes and nanosorbents (Nanosorbents contain high sorption capacity which carries many applications for water treatment methods) and some imprinted polymers are effectively useable for treatment process of contaminated water. In short, the study of nanomaterial applications in water purification is considered to assess positive perspectives [16]. Therefore, we have summarized the role of nanomaterials for wastewater treatment to overcome the water crises in this review article. Nanoengineered materials provide a potential and significant water treatment approaches which can be easily adaptable, but some imperfections still need for further attention which are specially summarized in this article. Moreover, we also addressed the limitations, advantages, disadvantages and future perspectives related to these nanomaterials. Furthermore, the toxicity of nanomaterials and their several applications in wastewater treatment are briefly discussed which might be useful for researcher to plan new strategies.

2. Water Treatment Methodologies

2.1. Nanophotocatalysts

The word "photocatalysis" is composed of two Greek words "photo" and "catalysis" which mean decomposition of compounds in the presence of light. Usually, in scientific world there is no consensus definition for photocatalysis [28]. However, this term can be employed to define a process to activate or stimulate the substance by using light (UV/Visible/Sunlight). Photocatalyst which changes the rate of reaction without any involvement by itself during the chemical transformation process. Furthermore, the key difference between traditional thermal catalyst and photocatalyst is that the prior is activated through heat while the final is activated through photons of light energy [29]. Nanophotocatalysts are commonly used for wastewater purification, as they help to enhance the reactivity of catalyst due to having a greater surface ratio and shape dependent features [30].

The nano size-based materials show different response as compared to bulk materials due to their distinct quantum effects and surface properties. It assists to increase their electric, mechanical, magnetic chemical reactivity and optical properties too [31]. It has been showed that the nanophotocatalysts can expand the oxidation ability due to effective production of oxidizing species at surface of material which helps in degradation of pollutants from the polluted water effectively [32].

Nanoparticles like zero-valence based metal, semiconductor and some bimetallic type etc. are mostly used for treatment of environmental pollutants e.g., azo dyes, Chlorpyrifos [33–35], organochlorine pesticides, nitroaromatics, etc. [36]. There are also several reports in literature which illustrated that TiO_2 based nanotubes can effectively be used in removal of pollutants (organic pollutants such as azo dyes, Congo red, phenol aromatic base pollutants, toluene, dichlorophenol trichlorobenzene, chlorinated ethene, etc.) from waste water [37–41]. However, the most common and significant metal oxide nanophotocatalyst are SiO_2, ZnO, TiO_2, Al_2O_3, etc. [42–44]. Among them, Titanium dioxide (TiO_2) is one of excellent photocatalyst from all existing material due to its several reasons such as low cost, toxic free property, chemical stability, and its easy availability on earth. Moreover, TiO_2 exist in three natural states, anatase, rutile, and brookite. So far, anatase considered as good nanophotocatalyst material [45]. The bandgap of this state is 3.2 eV and it can absorb ultraviolet light (below 387 nm) [46]. However, other photocatalysts like ZnO have also been produced to eliminate contaminants in wastewater and presents reusable ability effectively [47–51]. In case of composite nanomaterial, the degradation of reference substance (dimethyl sulfoxide) for evaluating the photocatalytic performance of water treatment by using CdS/TiO_2 composite as catalyst under the irradiation of visible light also investigated [52]. The iron doped nanomaterials have ferromagnetism ability which helps to recycle and reuse it easily [53–56]. Similarly, due to some characteristics like high photocatalytic reactivity Pd incorporated ZnO nanomaterial have been used for the removal of *Escherichia coli* from wastewater [57]. Although, new efforts have been targeted on metal oxides in order to increase the photocatalytic performance under visible light irradiation through modifying them with other elements like metals or metal ions [6,33] carbonaceous-based materials, dye sensitizers [58] and many others but still there is a need for further modifications in nanophotocatalysts.

Furthermore, nanophotocatalysis process may occur in two states heterogeneously or homogeneously. The most intensively studied state is heterogeneous nanophotocatalysis in modern era, due to its wide scope in water decontamination and environmental related applications. In case of heterogeneous photocatalysis, it implies the prior development of an interface between fluid (both reactants and products of reaction) and solid photocatalyst (such as metal or semiconductor) [59,60]. Generally, the word "heterogeneous photocatalysis" is mostly employed where a light-based semiconductor photocatalysts are used, which is in interaction with gaseous or liquid phase [61]. The heterogeneous photocatalysis based applications are strongly depend on the scaled-up reactor based on advance developed designs with improved efficiency [62]. The major task in reactor designing is effective illumination of nanocatalyst and mass transfer optimization, particularly in case of liquid phase. The transfer of photon can be improved by using light-emitting diodes and optical fibres, but productive revolutions in this field are still lacking. Moreover, an extensive effort has been focused toward the progress of solar photoreactors [63–65]. According to literature, the positive role of nanophotocatalyst has been demonstrated in research laboratory for air cleaning and water treatment. At the commercial level it is still not a perfect way to minimize the problem. Moreover, the present lack of extensive commercial applications is due to absence of effective photoreactor configurations and lower photocatalytic competence of photocatalysts. Despite all, heterogeneous nanophotocatalyst recommend fascinating advantages i.e., inexpensive usage of chemicals, additive free, work even at lower concentration, chemical stability (e.g., TiO_2 stable in aqueous medium) [66]. Therefore, recently heterogeneous photocatalysis is achieving the pre-industrial scale.

2.1.1. Advantage and Disadvantages of Nanophotocatalyst

Nanophotocatalysis has reflected a vital role for the mineralization of dangerous organic substances at 25 °C and proved very effective and efficient method for detoxification of water with help of nanophotocatalysts [67]. Furthermore, mostly nanophotocatalysts show some advantages such as they are less toxic, having low-cost, chemically stable, easily accessible and excellent photoactive properties with nano size i.e., 1–100 nm range [68]. Generally, nanophotocatalysts present various advantages such as stability (chemical/physical), low cost and eco-friendliness. Among them, TiO_2 having good photostability but many nanophotocatalysts, such as zinc oxide, metal sulfide materials, copper-based materials, and so on exhibit relatively low chemical stability due to photocorrosion [69]. As light shine on them, they oxidized or reduced depend on materials and their oxidation states changes by generating holes or electrons, which leads to decomposition of photocatalysts and decreases the efficiency of photocalatalysts. Therefore, there is a strong need to synthesize a nano composite in order to achieve stable photocatalysts for long-time performance.

The major advantage of nano-sized is related to quantum-size effect, which enhance the energy bandgap and reduce the particle size [70]. Moreover, as a process photodegradation also have several advantages such as low cost, reusable, and generally complete degradation. In addition to all of the developments, still nanophotocatalyst facing some issues i.e., toxicity and recovery of catalysts from mixture. These type of issues limits the applications and scope of nanophotocatalyst at higher level [71]. So, the scientific community is now focusing others nanocomposite of different material which can reduce the toxicity while using in water treatment process. So, it is suggestible for the scholar community to synthesize new photocatalysts which can work in visible ranges for sustainable result and promote the doped photocatalyst with different material such as graphene and its derivatives to reduce the toxicity effect. To overcome the catalyst recovery drawback, one significant approach could be used i.e., magnetic nanophotocatalysts in wastewater treatment. When magnetic nanophotocatalysts are used, the recovery of catalyst can be achieved through external magnetic fields, therefore, permitting the many recycling of nanocatalyst and achieved more effective and naturally responsive water decontamination processes. Furthermore, the approach regarding nanophotocatalysis for removal of pollutants from water has been described as a very effective approach and some unique applications of nanophotocatalysts are shown in Figure 1.

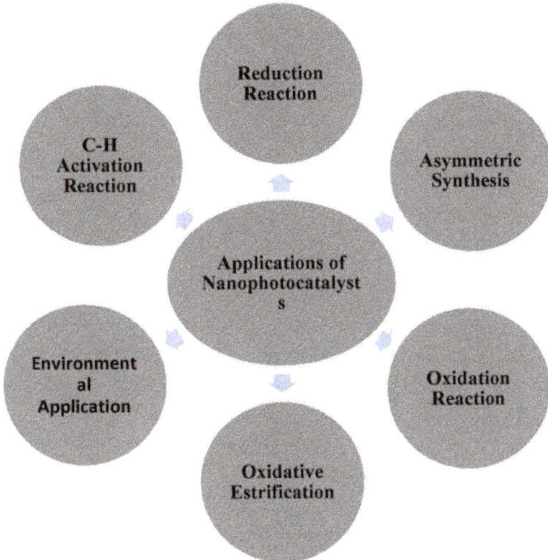

Figure 1. Potential applications of nanophotocatalysts.

2.1.2. Future Perspectives of Nanophotocatalyst

A wide research on nanomaterials is ongoing in field of nanophotocatalysis which led to few progress in reactor designing and developments regarding the modifications in nanophotocatalysts. Although many developments in nanophotocatalytic materials occurred, still some important inquiries required related to characteristics of nanophotocatalytic materials. The major challenges in intensification process is mass transfer limitations and higher consumption of photons [72]. The concept of nanocomposites is ideal in solving the electron pair recombination problem which can be prolonged by combining the nanocomposites with nanophotocatalytic reactor structures. The new modern reactors are known as microfluidic reactors which open a new door for intense characteristics study in reaction phase and synthesis phase. Microfluidic reactors are those reactors which are working on micro level with reactants [73,74]. The key features of micro-reactors are large surface-to-volume ratio, improved diffusion effect and great mass transfer coefficient factor, highly stable hydrodynamics, less Reynold's flow, and informal handling which makes them more ideal material than conventional reactors. However, still it is difficult for photocatalysis to apply on large scale in actual wastewater.

Moreover, the synthesis of significant structures such as nanorod, nanosphere, nanoflowers, nanoflakes and nanocones with enhanced functional and structural properties could be opened an extensive area of study. Several structures of nanomaterials with potential properties could be produced through the measured synthesis approaches. However, the future research, should be explored by producing new photocatalysts. The synthesis of novel nanophotocatalysts with excellent efficiency, inexpensive, eco-friendly and high stability is crucially needed. Moreover, to exploit pollutant treatment effectively, several approaches should be joint with sensible match, such as electrocatalysis, photocatalysis, adsorption and several thermodynamics processes. The preparation of nanocomposites in the presence of ZnO, TiO$_2$ was well explored in early decades for the treatment of water pollutants. The nanocomposites preparation by using carboneous material, polymer and ceramic materials are still in initial stage. It can generate ideal nanocomposites with improved properties. The heterogeneous photocatalytic for wastewater remediation is inhibited by some main technical problems that need to be study effectively. Finally, a significant photocatalytic treatment with better solar-driven, excellent efficacy and less site area requirements can be comprehended in future with fast assessment.

2.1.3. Photocatalytic Degradation and Mineralization Pathway

Nanomaterials also got much attention in degradation as well as mineralization of toxic organic pollutants due to having remarkable physiochemical properties. In photocatalysis, there are two types of processes that occur, namely mineralization and degradation of organic pollutants [5,75]. In the process of degradation, the organic pollutants are splitting or decomposed into several products while in the case of mineralization, the complete destruction of organic pollutant took place into water, carbon dioxide and some inorganic ions. The possible pollutant degradation mechanism in the presence of light is shown in Figure 2 [75]. Briefly, a semiconductor like TiO_2 absorbs the light which is greater or equal to TiO_2 band gap width, it carries to electron–hole pairs (e^-–h^+). If the separation of charge is continued, the electron–hole may travel to the surface of catalyst where they contribute with sorbed species in redox reactions. Particularly, h^+_{vb} react with water (surface-bound) to generate the hydroxyl radicals and simultaneously e^-_{cb} selected by oxygen to produce the radical anion (superoxide radicals) as designated below in equations.

$$TiO_2 + h\nu \rightarrow e^-_{cb} + h^+_{vb} \tag{1}$$

$$O_2 + e^-_{cb} \rightarrow O_2^- \tag{2}$$

$$H_2O + h^+_{vb} \rightarrow OH + H^+ \tag{3}$$

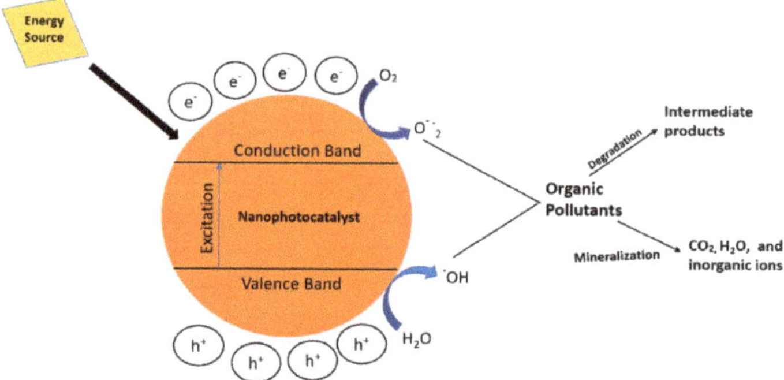

Figure 2. General mechanism of toxic organic compound degradation through nanophotocatalysts.

After pollutant degradation by using nanomaterials in the presence of light, HPLC-MS or GC-MS was employed to analysis the produced degraded products. Here, it is necessary to confirm regarding the obtained degradation products, whether these products are more toxic or less toxic as compared to parent compound using toxicity test.

As discussed above, mineralization concept is actually synonymous of complete photodegradation. It defines the degradation of a compound into CO_2 and H_2O. Sometimes others minerals also released during mineralization of compounds such as sulphate, ammonia, sulphite, fluoride, sulphide, chloride, phosphate, nitrite, etc. [76]. Generally, the rate of mineralization is less as compared to degradation, probably due to generation of stable intermediate during process. Therefore, it is supposed that a long irradiation is compulsory for entire removal of total organic carbon (TOC). Moreover, the mineralization concept is avoiding the generation of undesirable products is the excellent mode to degrade the organic compounds. The TOC is defined as the total quantity of bounded carbons in any organic compound and measured by TOC analyser. The possible mechanism of pollutant mineralization is shown in Figure 2 coupled with photodegradation mechanism.

For example, different degradation products as well as the mineralization products of Lignocaine are shown in Figure 3 as follows [77];

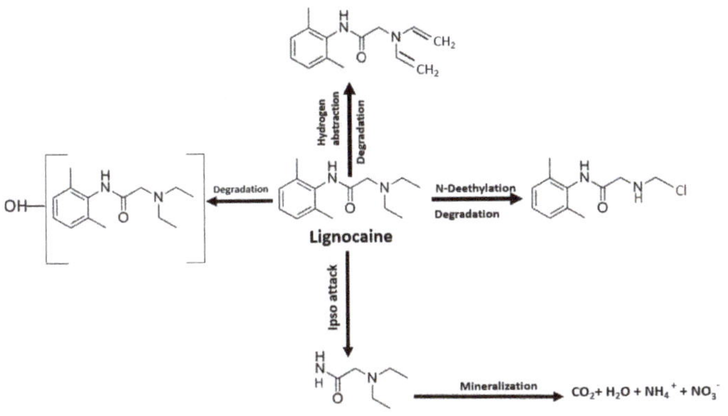

Figure 3. Different degradation and mineralization products of Lignocaine.

2.2. Nano- and Micromotors

Nanotechnology, an area of research that has progressive at such a quick pace in early decades and offered many approaches for water treatment. In modern era, nano/micromotors have been considered that can convert the energy from several resources into machine-driven force, empowering them to achieve special goals through various mechanisms. These innovative motors are motorized in both cases by using fuel or without fuel sources (acoustics, magnetic field, electric field) have several significant exciting applications [78]. They show more speed, high power, specific control movement, self-mix ability, etc. The removal of contaminant from polluted resources is a significant focus for environmental stability and sustainability [79]. The trend of water purification and its treatment has grown quickly throughout the world because of high demands for pure water resources and novel water superiority regulations. A large variety of approaches are used earlier for removal of pollutants from polluted groundwater, fresh water, sediments wastewater, etc. Different mechanisms employed by nano/micromotors for treatment of water pollutants are graphically shown in Figure 4 and some environmental applications of nano/micromotors are shown in Table 2 with their mechanism.

Traditional treatment Processes are inadequate through diffusion and, hence, entail outward agitation to stimulate the wastewater treatment process. Though, nano/micromotors could possibly overwhelmed the diffusion boundary by energetic mixing due to their self-propulsion competences. These self-propelled nano/micromotors expressively stimulate the water treatment efficiency through water decontamination efficiency, merging with materials nano/microstructure i.e., greater surface area and working activities [80–83]. Moreover, nano/micromotors can go in nano/microscale detentions in the presence of a magnetic field, serving as programmed cleaners, where conventional approaches are not actively working. However, most positive concept for nano/micromotors in term of wastewater decontamination depends on fuels as shown in Table 3. There is problem, this condition reduced the working potential of nanomotors in biological applications. Although some photocatalytic, biocatalytic and magnetically driven nano/micromotors have been industrialized, still there are some challenges in order to apply nano/micromotors for water treatment in future.

Table 2. List of nano/micromotors and their applications.

Motors	Working Mechanism	Applications	References
Zn, Al/Pd micromotors	speed-pH dependence	pH controlling	[85]
Hydrophobic agglomerates of pollutants	Surface tension induced	Increased diffusion of pollutants	[86]
Polymer capsule motors	Surface tension induced cargo towing	Oil remediation	[87]
Au/Pt nanomotors	silver-induced acceleration	Detection of silver ions	[88]
Ag-based Janus MIP microparticles	Molecular imprinted polymer recognition	solid-phase extraction	[89]
Au/Pt nanomotors	DNA hybridization through using Ag nanoparticle	DNA detection	[90]
Bubble-propelled Pt-based micro engines	High fluid transport	Oxidative detoxification of nerve agents	[91]
Bubble-propelled Pt-based micro engines	Fenton reaction; high diffusion	Degradation of organic pollutants	[92]
SAM modified Pt micro engines	Hydrophobic interactions with oil droplets	Oil removal	[93]
Ir/SiO$_2$ Janus motors	Speed-concentration dependence	Detection of hydrazine	[94]
Pd nanoparticle-containing Microspheres	pH dependence	pH monitoring	[95]

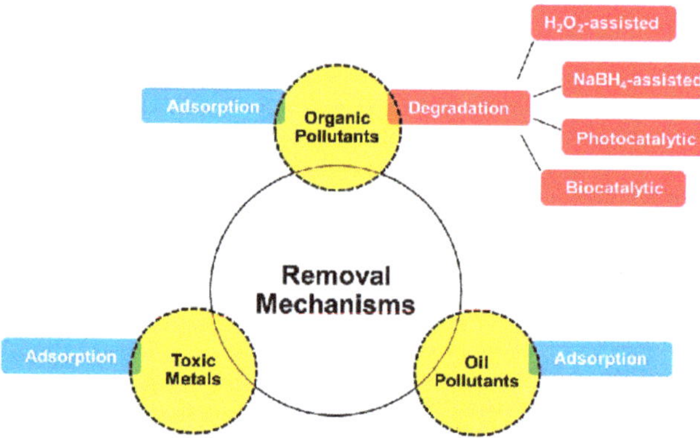

Figure 4. Different pollutant removal mechanisms used by nano/micromotors (adapted with permission from Royal Chemical Society) [84].

Table 3. List of nano/micromotors for water decontamination in presence of fuel.

Type of Nanomotors	Target Pollutants	Operational Mechanism	Fuel	References
PEDOT/Pt bilayer nanomotor	Organophosphorus nerve Agents	Oxidative neutralization	H_2O_2	[96]
Ag-incorporated zeolite	-	Adsorptive Detoxification	H_2O_2	[97]
rGO-SiO$_2$–Pt Janus magnetic micromotors	Polybrominated diphenyl ethers (PBDEs) and 5-chloro-2-(2,4-dichlorophene) phenol (triclosan)	Adsorption	H_2O_2	[98]
Pt coating Activated carbon-based motors	Heavy metals (Pb^{2+}), nitroaromatic explosives, dyes	Adsorption of active carbon	H_2O_2	[99]
Zero-valent-iron/platinum (ZVI/Pt)	Methylene blue	Fenton reaction	H_2O_2	[100]
CoNi@Pt nanorods	4-nitrophenol, Methylene Blue, and Rhodoamine B	Degradation	Borohydride	[82]
Pd-Ti/Fe/Cr tubular microjets	4-nitrophenol (4-NP)	4-nitrophenol (4-NP)	Borohydride	[101]
Au NPs/TiO$_2$/Pt nanomotor	super organic mixture	Photocatalytic Degradation	H_2O_2	[102]
Polystyrene–Zn–Fe coreshell Microparticles	Rhodamine B	Fenton reaction	H_2O_2	[103]
TiO$_2$/Au/Mg microspheres	Mineralization of the highly persistent organophosphate nerve agents, bis (4-nitrophenyl) phosphate (b-NPP) and methyl paraoxon (MP)	Photocatalytic degradation	-	[104]
Biotin-functionalized Janus silica Micromotor	Rhodamine B	Charge adsorption	H_2O_2	[105]
Fe^0 Janus nanomotors	Azo dyes	Fenton reaction	Citric acid	[106]
CoNi-Bi$_2$O$_3$/BiOCl-based hybrid Microrobots	Rhodamine B	Photocatalytic degradation	UV light	[107]
DNA-functionalized Au/Pt Microtubes	Hg^{2+}	Adsorption	H_2O_2	[108]
GOx-Ni/Pt	Pb^{2+}	Adsorption	H_2O_2	[109]
3D printed motors (TSM)	Oil droplets	Adsorption	-	[110]
Alkanethiols-coated Au/Ni/PEDOT/Pt microsubmarine	Oil droplets	hydrophobicity upon the oil-nanomotor interaction	H_2O_2	[111]

2.2.1. Advantage and Disadvantages of Nano- and Micromotors

Recently, nano/micromotors are used to overcome the environmental issues such as water pollutant treatment and environmental sensing/monitoring. The nano/micromotor are considered as excellent option, because the reactive nano-based materials have potential properties which make it more efficient to convert toxic pollutants into toxic free. Nanomachines offer different advantages as compare to traditional remediation agents. The small machines improve a dimension which based on decontamination approaches, lead to in-situ and ex-situ nano-remediation rules, and have ability to decrease the clear-out time and entire cost [112]. Specifically, the constant movement of nanoscale substances can be utilized for transferring reactive nanomaterials for water purification through polluted samples, for discharging remediation agents to long distances, and for communication of important mixing throughout decontamination processes [113]. Existing technologies are insufficient for fulfilling the demand in term of scaling up, as stated in the treatment of polluted water, therefore more efforts are required. Some issues require to be explained prior to move toward applied application at commercial level. For example, the lifecycle of multi-functional nano/micromotors is restricted to the residual materials in its physique that were consumed in locomotion or oxidation reactions. Another disadvantage is poisoning of Pt layer because the compounds present in wastewater can bond chemically to other surface-active sites of the catalyst, or great viscosity-based wastewater which hinder the movement of micromotors [114]. The introduction of novel development in nano/micromotors will offer countless environmental treatment possibilities to achieve more multifaceted and challenging operations.

2.2.2. Future Perspectives of Nano- and Micromotors

Nano/micromotor are still considered as immature techniques because it still on initial stages but this topic open a new door of research for researcher. As Compared with other traditional wastewater treatment approaches, this subject is still novel and but has a lot of restrictions in extensive and real-world applications. Despite all, current approaches are insufficient to fulfil the current demand in scaling up the technique described in water pollutant treatment, therefore considerable work is still required [115]. Novel materials should link with nano/micro-motors, such as graphene to treat wastewater. These kinds of materials show better performance in wastewater treatment as compared with other materials. Moreover, some new mechanism should functionalize in nanomotor design for better results. Soon, we believe that excellent nanomotor will be planned by associating with other approaches. These predictable new, novel micro/nanomotors would ultimately develop the ecological monitoring and wastewater treatment technologies, from a struggle to advance the class of life.

2.3. Nanomembranes

Nanomembranes are a unique type of membrane formed with different nanofibers which are employed to eliminate unwanted nanoparticles present in aqueous phase. Using this technique, the process takes places at a very high elimination speed with condensed fouling propensity and it also served as a pre-treatment process which is used for reverse osmosis [116]. There are many reported studies on membrane nanotechnology in order to produce multifunction membrane by using different nanomaterial substance in different polymers-based membranes. Water porous membrane for water treatment did nanofiltration, ultrafiltration, reverse osmosis, etc. The membrane contains a porous support with composite layer. Typically, the considerable composite layer is carbon-based material (graphene oxide/CNT) dispersed into polymer matrix for significant practice. This provide promising and significant progresses in fouling resistance and aqueous transport. For example, CNTs hold anti-microbial properties that can minimize fouling, biofilm formation and it can also reduce the chance of mechanical failures [117]. The doping process of nanomaterial (zeolite, alumina, TiO_2, etc.) into polymer ultra-filtration membranes show the formation of amplified membrane on surface of hydrophilicity and fouling resistance [118–120]. The antimicrobial material like silver metal particles are also doped with a polymer to produce polymeric membrane to prevent attachment of bacteria and inhibit biofilm production on surface of membrane [121,122]. It served to inactivate viruses and it has also ability to prevent the biofouling of membrane [123]. The production and growth of nanomaterial thin film are incorporated into active thin film composite through doping in surface modification. Usually, there are a few major challenges in nanomembrane i.e., membrane clogging and membrane fouling. Therefore, it is a need to overcome this problem. By the addition of super hydrophilic nanoparticles in making a thin film nanocomposite membrane, the prevention from clogging and fouling could be checked. Furthermore, metal oxide nanoparticles (Al_2O_3, TiO_2), antimicrobial nanoparticles (nano silver, CNTs) and aquaporin-based membranes are very useful material to overcome the membrane clogging and fouling issue due to having high hydrophilicity, high porosity, better fouling resistance and a better homogenous nanopore. Generally, nanoparticles affect the selectivity and permeability of a membrane which depends upon the sort, quantity, dimension, etc. of nanoparticles. Moreover, there are many biological membranes present with highly permeable and selective ability [124,125] Nanomembranes are also used for wastewater treatment due to having several properties that make this material more prolific, these are high uniformity, homogeneity ability, optimization, short time required, easily handled and contain much order of reaction [126]. There are some nano photocatalysts which can be introduced in the nanocomposite membrane to make fit it for the degradation of organic pollutants. For example, TiO_2 incorporated nanomembranes and films are effectively used to deactivate different microorganisms and degrade the organic pollutants [127]. The developments and progressive growth of nanotechnology especially in nanomaterial produced several nanostructured catalytic membranes which contain novel properties like improved selectivity, high decomposition rate, and larger foul resistance [128,129]. In order to synthesize these types of nanomaterials, there were many approaches used to produce it with multi-functionality features [130]. The incorporation of nanostructured catalytic materials like iron-catalysed based free radical and enzymatic catalysis within pore membrane showed a constructive progress in this technology. Therefore, it possible to carry out oxidative reactions for removal of pollutants and detoxification of water without use of any toxic chemicals. To prove the efficiency of nanostructured catalytic membrane in industrial/commercial applications, immobilized membrane nanoparticle (ferrihydrite/iron oxide) reaction was carried out with hydrogen peroxide to make free radicals' ions for removal of chlorinated based organic pollutant in real groundwater. The development of nanostructured materials is still very useful in many other environmental applications [131]. There were several reported studies of metallic nanoparticles immobilization on membrane (e.g., chitosan, cellulose acetate, polysulfones, etc.) for dichlorination, degradation of a toxic substance which contains novel properties such as, hindrance of nanoparticles, high reactivity, absence of agglomeration and surface reduction [132,133]. Palladium acetate and polyetherimide both are used to prepare nanocomposite films and a novel type interaction is present among hydrogen and Pd nanoparticles in order to improve water treatment efficiency [134]. In situ and ex-situ methods are also used to produce nanomaterials within the matrix with precursor film under many conditions [135–137]. This offers many opportunities to produce nanomaterial with tunable properties. Moreover, the developments of nanotechnology are also responsible to produce many significant catalytic membranes with high selectivity, better permeability and high resistance fouling. Modern methods contain hybrid and bottom-up approaches for empowering its multi-functionality characteristics in the field of wastewater treatment [138].

2.3.1. Advantage and Disadvantages of Nanomembrane

The main objective of using membranes in case of drinking water is to separate the toxic particles from water resources. The nanomembrane filters were also used to measure the water safety level [139]. The advantages of nano membrane as compared to traditional approaches for filtration is that; in traditional approaches of filtration, throughout the entire process, calcium and magnesium required another ion to compensate, therefore mostly Na^+ ions are served as an exchanger but in case of nanomembrane, there will be no need [139,140]. Nanomembranes limitations are usually reducing its efficiency compared to other conventional approaches. The first one is fouling of nanomembrane which comes after using the membrane few times. It is the major issue, which make this approach more expensive and inefficient. Fouling problems occur in the nanomembrane because it entirely depends on working conditions, sometime the working conditions are not appropriate such as over temperature, pressure, and optimization is also considered as responsible for membrane fouling [141]. Second major limitation is membrane stability. The nanomembrane could not keep the stability for long period. After sometimes, the stability start reducing its efficiency and it does not give excellent outcomes as earlier. So, there will be a need for changing the nanomembrane to get excellent results, but this will cause several other issues such as high cost, impurity chances during changing process, etc. [142]. The stability is depending on the essential chemical resistance which apply for material cleansing. The disadvantages are less reliability, slow operation process, less selectivity, high maintenance cost and working efficiency reduce with passage of time. There are some common disadvantages that is why it is not extensively explored [143]. Some nanomembranes types are summarized in Table 4 along with their advantage and disadvantage and applications.

Table 4. Advantages/disadvantages and applications of nanomembranes.

Different Type of Nanomembranes	Advantage of Nanomembrane	Disadvantages of Nanomembrane	Application	Reference
Nanofiber membranes	Excellent porosity, tailor-made, bactericidal, good permeate efficiency,	Pore blocking, conceivably discharge of nanofibers	Ultrafiltration, prefiltration, filter cartridge, water handling, separate filtration devices	[144]
Nanocomposite membranes	High hydrophilicity, better water permeability, high fouling resistance, good thermal and mechanical stability	Resistant substance substantial required when oxidizing nanomaterial, used to discharge nanoparticles	Entirely dependent on composites	[145]
Aquaporin-based membranes	Improved ionic selectivity and better permeability	Poor mechanical stability	Less pressure desalination	[146]
Self-assembling membranes	Homogeneous nanopores membranes	Laboratory scale availability only	Ultrafiltration	[147]
Nanofiltration membranes	Charge-based repulsion, comparative less pressure, better selectivity	Membrane blocking	Colour, Reduction of hardness, odour	[148]

2.3.2. Future Perspectives of Nanomembranes

Nanomembranes separate the inorganic ions, organic, nanoparticles viruses-based pollutants from water resources through solution using diffusion and size exclusion. Though these described nanomembranes are proved effectively at the research laboratory level, upscaling to lower cost, but making them ideal for industrial scale is still emerging as a challenge. To cross this barrier toward productive upscaling, the commercialization objective needs a joint collaboration struggle through research institutions and manufacturing companies. The selectivity of nanomembrane should be improved, secondly upgrade the nanomembrane resistivity to avoid membrane fouling. Surface grafting-based polymers like zwitterionic may be a significant candidate for the development of new generation membranes. Surface grafting, though, would not report fouling resistance to inner walls pore [149]. Thereafter, the prime of other appropriate anti-fouling convertors fixed into the membrane matrix that will be vital. It is also very significant to improve the high sensitivity of polyamide membranes to many types of oxidants like ozone and chlorine. Furthermore, the multi-functional membranes fabrication requires a significant attention for better innovation at industrial scale. However, still there are several challenges to be solve for manufacturing production of lower cost and effective nanomembranes for water treatment.

2.4. Nanosorbents

Nanosorbents hold wide properties like high sorption capability that make the nanosorbents more appropriate and powerful for water treatment [150]. These nanosorbents are very rare in commercial form but researcher and experts doing a lot of work on it to produce nanosorbents in larger quantity/at commercial level [151]. The most commonly reported nanosorbents are based on carbon (e.g., carbon black, graphite, graphene oxide). Furthermore, metal/metal oxide and polymeric nanosorbents were also exist [152]. The composite of different material like Ag/polyaniline, Ag/carbon, C/TiO_2, etc. carrying huge significant importance in order to reduce the effect of toxicity in the wastewater treatment process. The carboneous material such as CNTs with a cylindrical form nano structure may present as single-walled and multiwalled nanotubes depending on the method of synthesizes. CNTs hold measurable adsorption sites and due to high surface area, they hold sustainable surfaces. It must be stabilized to prevent from aggregation that decreases the surface-active sites because CNTs has hydrophobic surface properties. So, it is an efficient material for the pollutants by adsorbing process. Similarly, the polymeric nanoadsorbents like dendrimers are functional for eliminating organic pollutants and heavy metals from wastewater [153]. For example, copper ions were reduced with the help of dendrimer-ultrafiltration system [154]. They simply regenerated by shifting of pH effect and show biocompatibility, biodegradability, and toxic free environment. Furthermore, the removal percent of dyes or others organic pollutants is almost 99% [155]. Another important nanosorbent is zeolites, which have an absorbent structure in which several nanoparticles like copper, silver ions can be implantable [156]. The advantage of zeolites is to control the amount of metals and it also served as anti-microbial agent [157]. Moreover, the magnetic nanosorbents have play vital role in water treatment and a unique tool to remove different organic pollutants from water [158]. Some organic containments are also removed by using magnetic filtration. Magnetic separation nanosorbents are synthesized by ligands coating with magnetic nanoparticles at specific affinity [159]. There were many methods like ion exchange, cleaning agents, magnetic forces, etc. reported to regenerate these nanosorbents. These regenerated nanosorbents have the ability of being cost-effective and more promotions of commercialization. Despite all, the carbon has some health risks. The reported studies demonstrated that toxic effect is depends on morphology of nanoadsorbents, chemical stabilizers and surface modifications [160,161]. So, there is a need to give an attention to synthesize more stable morphology (size and shape) to overcome the toxicity issues as well as health risks. Furthermore, the bioadsorbents have the properties of high biodegradable, good biocompatible and nontoxicity which could be replaceable with chemically synthesized nanosorbents. The graphene oxide is suggestable to scientific community because it is very emerging nanomaterial to use as nanosorbents to remove pollutant and it can give better result than others due to its superior properties. There are some reported nanosorbents and their functions as shown in Table 5.

Table 5. Most commonly used nanosorbents and their functions.

Different Nanosorbents	Treatment Function	References
Carbon-based nanosorbents	Nickel ions present in water	[162]
Graphite Oxide	Removal of dyes	[163]
Regenerable polymeric Nano sorbents	Organic pollutants, inorganic contaminants of wastewaters	[164]
Nanoclay	Hydrocarbons, Dyes	[165]
Nano-metal oxides	Different Heavy metals	[166]
Nano-Aerogels	To remove the uranium fromdrinking water	[167]
Nano-iron oxides	To eliminate the hormones andtoxic pharmaceuticals materialfrom water	[168]
Polymer Fibers	To remove the arsenic and other toxic metals	[169]

Advantages, Disadvantages and Future Perspectives of Nanosorbents

The role of nanomaterials as sorbents in explaining ecological issues such as decontamination of wastewater received a great interest due to having remarkable physiochemical properties. These properties distinguish them in several fields as compare to conventional traditional sorbents. For an ideal sorbent to treat the pollutant very effectively in a short time, it should hold large surface area, excellent rate of adsorption, short time adsorption and equilibrium times. Similar to nanosorbents, nanomaterials got interest because of having nano-size which can hold excellent rate of adsorption with short time. Furthermore, nanosorbents that can be used as a separation medium in water decontamination to eliminate the organic, inorganic-based pollutants from polluted water are nanoparticles [170]. Literature review shows that a lot of efforts are already done for wastewater treatment and used nanomaterials as sorbent for efficient results. Some challenges are required to be addressed entirely for commercialisation purpose of the nano-size sorbents for water decontamination such as production scalability, as well as excellent adsorption measurements, selectivity, stability of material, operational duration of material, etc. However, there is an enormous need for an active approach to treat the wastewater and fabricate some new nanosorbents which could be applied to control toxic ions and compounds from wastewater [171]. The future prospects looking admirable, as scientific community working on advances knowledge and improving the adsorption mechanisms. In the modern world, researchers, consultants and officials are concerned about wastewater pollutant which holds health risks and they all are dedicated to find an effective explanation as concerning the nanomaterials should be used at industrial scale.

3. Self-Toxicity of Nanomaterials

Nowadays, nanomaterials have become most attractive and widely used material for different applications in various fields such as electronics, medicine, agriculture, wastewater treatment plants, energy generations and other sciences. These nanomaterials are doped metal oxides, doped carbon nanotubes (single and double walled), nanosorbents, etc. There were many reports available in literature which show that these nanomaterials served as a leading role in order to remove different pollutants from waste waters [172]. In addition to their importance, utilities and applications in different fields, especially in water treatment process, it is very important to know about their self-toxicity effects. There are studies which show the toxicity effect of few nanomaterials, as shown in Table 6. Moreover, few metal oxides show some toxic character at high and some, even at low concentrations [173,174]. Furthermore, the toxicity of CNTs is depending on different properties like length, surface area, distribution ratio, aggregation degree, initial concentration of material [175]. Furthermore, single walled CNTs are less toxic as compared to double walled [176]. They are associated for pulmonary inflammation, oxidative stress, granuloma in lungs, basic inflammation, apoptosis and fibrosis [177]. Similarly, in case of TiO_2, the toxicities of different composition ratio of TiO_2 depend upon initial concentration and time. Generally, it is considered as non-toxic even at a higher concentration for 24 h [178]. Currently, the studies by scientific communities on these nanomaterials are ongoing to see the toxic and side effects and try to find out the mechanisms with a focus on explaining the outlines of nanomaterials transport, degradation, elimination, accumulation, etc. Furthermore, nanomaterials can harm body through various sources. Therefore, there is a critical need to find out the effects of nanoparticles on health.

Table 6. Observation of self-toxic effect of some common nanomaterials.

Nanomaterials	Observations	References
Gold nanorods doped poly(styrenesulfonate) (PSS) composite	Non-toxic	[177]
TiO_2, Al_2O_3/Carbon black Composite	More toxic at micron particle sized	[177]
Hexadecylcetyltrimethylammoni um bromide (CTAB) doped Au nano rode composite	High toxic at certain Concentration	[178]
Fe_2O_3 and carbon nanotubes Composite	It showed toxic effects and damage DNA even at lowest concentration	[179]
Carbon, metal, Al_2O_3 composite	Concentration- and time-dependent.	[180]
CdSe Quantum dots doped with polyvinylcarbazole composite	Acute toxicity observed	[181]
Single and Multi-walled carbon Nanotubes	Toxicity increases when concentration rises beyond 15 µg/cm.	[182]
Au/Carbon composite	Non-toxic at lower range.	[183]
Ag/Carbon composite	Time- and dose dependent but toxic	[179]

4. Applications of Nanomaterials

There were many opportunities to use engineered/modified nanomaterial for water treatment and many other industrial fields. Currently, nanomaterial attracts more attention of researcher in the field of wastewater treatment. The study trend in the field of nanomaterials is increasing day by day as we can observe through previously reported data. The number of publications in nanomaterial was less than 5% in 2005 but in 2019 number of reported works is more than 80% as shown in Figure 5 [184].

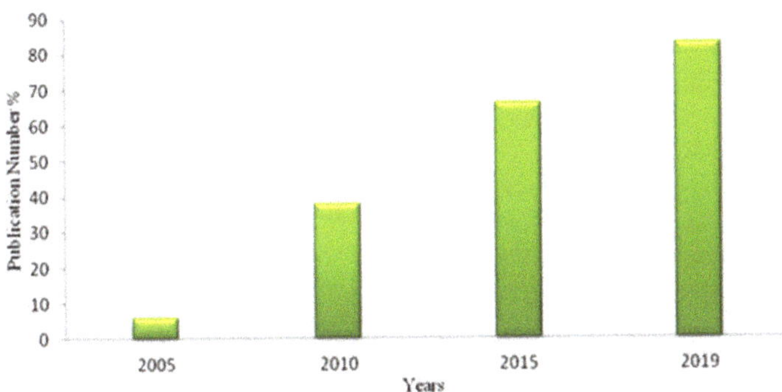

Figure 5. Publication trends in field of nanomaterial for wastewater treatment.

With emerging several aspects of nanomaterials, the wider environmental water resources effects are also considered. Such attentions might contain models to measure the significant advantages of reduction or inhibition of contaminants from engineering sources. Nanoscience expertise holds excellent potential for constant improvement for environmental protection. The present article has presented more indication to this matter and it has a response to what are the significant environmental impacts of this nanotechnology. For a quick assessment, the summary of nanomaterial applications for treatment of pollutants from water resources are briefly summarized in Table 7 [185].

Table 7. Summary of water treatments by using nanomaterial by different mechanisms [185].

Nanoparticles (NPs)	Target Analyte	Treatment Mechanism	Limitation	Positive Aspect
TiO_2	Organic Pollutants	Photocatalysis	Higher operation cost Tough to recovery, sludge production	toxic less, Water insolubility, photostability
Fe	Heavy metals, anions, organic pollutants	Reduction, adsorption	Tough to recovery, sludge Production, difficult sludge disposal, Health risk	In situ water remediation, Less cost, harmless to handle
Bimetallic NPs	Dichlorination	Reduction, adsorption	Tough to recovery, sludge Production	Higher reactivity
Nanofiltration and nanomembranes	Organic and inorganic substances	Nanofiltration	High cost, membrane Fouling	Low pressure
Magnetite NPs	Heavy metals, organic Compounds	Adsorption	Outside magnetically field needed for Separation	Easy separation, no sludge production
Metal-sorbing vesicles	Heavy metals	Adsorption	Re-use option, higher selective uptake profile, better metal affinity	Difficult to maintain stability
Micelles	Organic pollutants	Adsorption	In situ treatment, excellent affinity for hydrophobic	Costly
Dendrimers	Heavy metals, organic Pollutants	Encapsulation	Easy separation, renewable, high binding no sludge production,	Costly
Nanotube	Heavy metals, anions, organic pollutants	Adsorption	Dealing with pollution from water, Good mechanical properties, exclusive Electrical properties, Good chemical stability	High cost, lower adsorption process, Tough to recovery, sludge production, Dangerous Health risk
Nanoclay	Heavy metals, anions, organic pollutants	Adsorption	Lower cost, Exclusive structures, long stability, recycle, Higher sorption capacity, Informal recovery, better surface and pore volume	Sludge production

(1). Nanomaterials are very efficient in removing arsenic from drinking water when titanium nanoparticles and exchange based resin material impregnated with iron hydroxide material. The researcher also studied that titanium served as an adsorbent to remove arsenic from water present in packed bed reactor setting. In some developed countries, iron oxide is used as coated sand to remove arsenic from drinking water [186].

(2). Different nanomaterial like magnetic and carbon nanotubes can be served as sensor components due to having remarkable physical, electrical and chemical properties. Therefore, these sensors may offer opportunity to monitor water quality. Nanomaterial-based sensor is used to detect different pollutants because there have optical properties that make sensor more selective and sensitive to detect the pollutants [187].

(3). Recently, WHO reported that almost 783 million populations are suffering from fresh drinking water. Boschi-Pinto et al. stated that children deaths ratio of nearly 1.87 million is only due to water diseases [188]. Few conventional techniques are not suitable in underdeveloped countries because high investment is required for this project for maintenance expenses, and many other problems. Furthermore, generally people carry their water from another area and then save it for many days because there is no proper supply of drinking water. During collection and transport of water, there are chances of water contaminations. An efficient method of ceramic water filters (CWFs) provides an offer to treat these kinds of water infections (pathogens). Recently, CWFs are equipped by firing and pressing of flammable material and clay with silver nanoparticles [189]. The filter is produced by pressure and then dried air fired in a kiln. Therefore, a ceramic material produces filtered or clean water but by using this process removal is not possible up to high level. Silver doped nanomaterials could be used to get a higher percentage of pathogen's removal from water due to having antimicrobial properties. Ag solutions are used by brushing on CWF. It was observed that 80% of CWFs industries used Ag nanoparticles doped with some other materials. The Ag nanoparticles give remarkable results in order to remove different pathogens from drinking waters [190].

(4). Water pollutants are responsible for environmental pollutions. The polymer-based nanomaterial helps in environmental protection. Reported studies show that the purification of water using polymer material could be attained by nanoclay incorporation. The hydrophobicity enhancement helps to promote nanocomposite properties. Applications in contact with moist environments clearly indicate benefits from nanomaterial incorporation of nanoclay particles [191].

(5). The research is on the way to produce nanosorbents for different metals and organic compounds. Nanomaterial can act as sorbents like carbon nanotubes, zeolites and self-assembly layers on mesoporous supports, which control mesoporous ceramics with a sorbent that indicate efficient removal process of metals and anions from drinking water [192].

(6). Some nanomaterial has high antimicrobial activity properties. This type of material includes AgNPs, fullerene, TiO_2, CNT based nanocomposite, etc. and they contain several properties like mild oxidant, inertness to water and produce safe by-products [193].

(7). Now a days, many pollutants are present in water resources such as organics-based substances and even they present in trace amount which are very dangerous for human health. Usually, the chlorination and flocculation processes are used for removing water pollutants. There are some drawbacks in these conventional filtrations processes. These kinds of systems show less efficiency to remove pollutants entirely and also produces some sludge in water recourses which also creates big issues in environment pollution [194]. Therefore, to keep safe from sludge nanomembranes, it was employed because it does not allow any solid particle to pass into water and reduce the chances of sludge production in water resources. This reason makes nanomembrane more prominent in market than other methods. Certainly, some improvements are required to make it completely perfect technique.

(8). Metal oxides nanoparticles also play a very good role as catalyst in different oxidation reactions. They show strong catalytic reactivity towards pollutant molecules and change these pollutants into environmentally suitable products [195,196]. Some unique properties are present of these nanomaterials like nano size, high reactivity and greater surface area. Specially, TiO_2 photocatalysis plays a vital role in removing different impurities from surface water as shown in Figure 6.

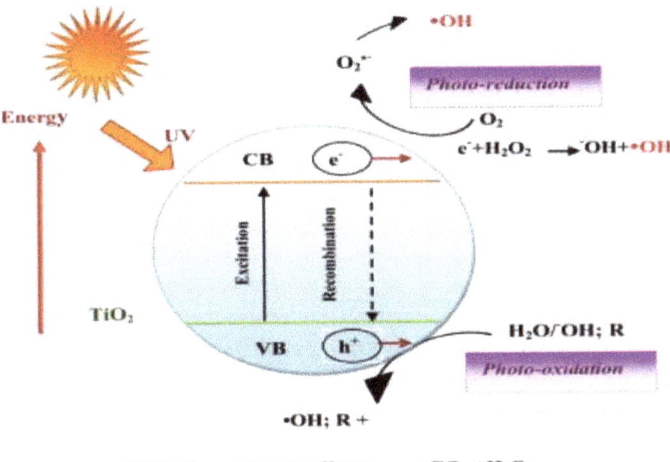

Figure 6. Schematic illustrating of TiO_2 photocatalytic process (Adapted with permission from Jurnal Teknologi) [197].

The summary/mechanism of pollutant removal by using TiO_2 nanoparticles as nanophotocatalyst is shown in Table 8.

Table 8. Observation of different pollutant removal by TiO_2 nanophotocatalyst.

Nanophotocatalyst	Target Analyte	Initial Concentration of Pollutant	Remediation Efficiency	Doses of Nanophotocatalyst	References
TiO_2	Nitrobenzene	50 mg/L	100	0.1M	[185]
TiO_2	Methyl orange	30 mg/L	100	3 g/L	[185]
TiO_2	Rhodamine 6G	125 mmol/L	90	0.1%(*w/w*)	[185]
TiO_2	Parathion	50 mg/L	70	1000 mg/L	[198]
TiO_2	Benzene	45 mg/L	72	5 g	[199]
TiO_2	Phenol	-	100	1.8 g/L	[185]
TiO_2	Rhodamine B	1.0×10^{-5} M	97	50 mg/50ml	[185]
TiO_2	Toluene	45 mg/L	71	5 g	[185]
TiO_2	Basic dye	20 mg/L	80	1.22 g/L	[200]
TiO_2	4-chlorophenol	1.0×10^{-5} M	99	25 mg/100ml	[201]
TiO_2	Procion Red MX-5B	10 mg/L	98	(2.0mg) TiO_2 (30%)	[202]

(9).   Different types of pollutants such as organic pesticides, organic dyes, pharmaceutical drugs, etc. are photodegraded by many researchers under several conditions such as choice of UV or Visible light, doped nanoparticles or undoped, metal/non-metal doping, etc. According to literature, as shown in Table 9 indicate clearly that modified or doped form of TiO_2 can give better results especially in photodegradation of pollutants. Table also demonstrates that the efficiency of doped-TiO_2 in visible light showed better results as compared to UV light.

Table 9. Observation of different pollutant removal by Doped-TiO_2 nanophotocatalyst.

Nanophotocatalyst	Target Analyte	Doping Agent	Remediation Efficiency	Source of Light	References
TiO_2	Glyphosate	Mn	80	Visible	[134]
TiO_2	Methylene Blue	Mn	75	Visible	[33]
TiO_2	Methylene Orange	Cu	100	Visible	[203]
TiO_2	Methylene Blue	S, I	90	Visible	[204]
TiO_2	Formaldehyde	N, S	65	Sunlight	[204]
TiO_2	Ramazol Brilliant Blue	La	72	Visible	[204]
TiO_2	Gentian violet	Mn	84	Visible	[194]
TiO_2	Acid Red 88	Mo	77	Visible	[194]
TiO_2	Rhodamine B	P	93	Sunlight	[205]
TiO_2	Methylene Blue	Fe	72	UV	[205]
TiO_2	Methylene Orange	Fe	99	UV	[205]

5. Nanomaterial Challenges for Water Treatment

Currently, the emerging nanomaterial possesses some challenges in the field of wastewater treatment [206,207]. Nanomaterial provides several possibilities of treatment of wastewater and they contain different kinds of substances which are distinct on the basis of the particles morphology. The developments regarding the commercial applications of nanomaterial is too quick and production of nanomaterial is increasing at a global level [58,208]. Nanomaterials are used for purification of polluted water through different methods such as photocatalytic, adsorption, and nanosorbents [209,210]. These methods need some modifications to work more effectively. Moreover, understanding of risk pretended by nanotechnology has not improved as quickly as research has giving possibilities to different applications of nanomaterials. The major challenge is the lack of information about the nanomaterial and how nanomaterials are released into environment, how they travel into water, how they start to exist in water [196]. Another challenge is related to human health because these types of material have some adverse effects/toxicological effect. The reported studies show that nanomaterials may cause some health issues but still research is going on. It is not easy to give a conclusion and ongoing research requires some conducting experiments at appropriate concentrations and more about toxicological studies. The membrane process is also effectively used wastewater treatment but there were very few reported studied about this process. The major challenge is the use of the membrane for fouling process and water treatment because after performing work/filtration, their pores may block, and the efficiency starts to decrease. In case of nanosorbents, the reusability is the main drawback, so there is need to synthesize such types of nanomaterials which can efficiently remove the pollutants from wastewater and after that the recovery process of these nanosorbents should be very easy. The United State Environmental Protection Agency (USEPA) found some basic challenges in order to remove nanoparticles using a process of water treatment [210–213]. (a) What is the mechanism of removing nanoparticles from wastewater? (b) What is the effect of nanoparticles on other waste substances during wastewater treatment? (c) How coagulation, carbon adsorption, etc. methods are effective in working. Moreover, water contains some pollutants and after treatment by the above mentioned methods, they degrade or are decomposed. There is a basic need to design an experiment for finding the intermediate products during these processes. Furthermore, constant development in methodology is needed to evaluate nanomaterials with low cost and it should be appropriate for complex nanomaterials.

6. Conclusive Remarks and Future Perspectives

Water distinguishes and makes our planet superior as compared to other planets. Though the worldwide available supply of pure water is high to meet all existing and predictable water demands. There are several areas where the drinking water resources are insufficient to fulfil the basic, economic and domestic developmental needs. In such areas, the insufficient fresh water to fulfil human water need and sanitation requirements is certainly a limit on human health and for other living creatures. Scientific community/research institutes must find a path to eliminate these limitations. Moreover, the world is facing several challenges in doing that, especially given a fluctuating and undefined future environment, a fast-rising population that is driving enlarged community and financial growth, urbanization and globalization. How superlatively overcomes on these challenges which entails exploration in all features of water management. The trend of nanomaterial for water pollutant treatment is rapidly increasing in this modern era due to very horrible conditions of water and demand for fresh water in the whole world. There is an important need for innovative progressive water treatment approaches, in specific to certify a high class of water for drinking purpose, remove micro/macro pollutants and increase industrial production developments through flexibly modifiable water treatment approaches. Nanotechnology has proved great achievement for controlling water purification challenges and makes some future advancement. Nanomaterial approaches like nanosorbents, nanostructured catalytic membranes, etc. are very efficient, less time required, less energy and eco-friendly techniques but all these methods are not cheap, and they are not used yet for commercial purpose to purify the wastewater at a large scale.

Nanomaterials show high efficiency due to having high rate of reaction. However, there are still some weaknesses that must be negotiated. Up to now, no operational digital monitoring techniques exist that offer consistent real-time measurement facts on the superiority of nanoparticles which are existing in small amounts in H_2O [212]. Furthermore, to reduce the health risk, some research institutes and international research communities should prepare proper guidelines to overcome this issue. Another, further mechanical restriction of nano-engineered water approach is that they are infrequently flexible to mass developments, and at present-day, in several cases are not modest with conservative treatment approaches [213]. However, nano-engineered materials provide excessive potential for water revolutions, in specific for decentralized water treatment technologies, point-of-use strategies, and seriously degradable pollutants. Furthermore, there is a great need to synthesize some modified nanomaterials which should be effective, having high efficiency, easy to handle and eco-friendly. It is also necessary to take the cost challenges and commercialization of these technologies for wastewater treatment. The different applications of nanomaterial can provide a tremendous offer in order to supply drinking water to whole world.

Author Contributions: A.A.Y., T.P. designed and wrote the first draft of review and drew the figures. K.U. participated in technical check and edited the full manuscript. M.N.M.I. was responsible for conceptualization of draft, technical check, and funding source for the manuscript. All authors discussed the results and commented on the manuscript. All authors have read and agreed to the published version of the manuscript.

Funding: This research was funded by Universiti Sains Malaysia, (11800 Penang, Malaysia). The grant number is 1001/ PKIMIA/8011070, and APC was funded by Universiti Sains Malaysia.

Acknowledgments: This research article was financially supported by Universiti Sains Malaysia, 11800 Penang Malaysia under the Research University Grant; 1001/ PKIMIA/8011070). The author (Khalid Umar) gratefully acknowledged the post-doctoral financial support. (USM/PPSK/FPD(BW)2/(2019).

Conflicts of Interest: The authors declared that they have no conflicts of interest.

References

1. Ahmad, A.; Mohd-Setapar, S.H.; Chuong, C.S.; Khatoon, A.; Wani, W.A.; Kumar, R.; Rafatullah, M. Recent advances in new generation dye removal technologies: Novel search for approaches to reprocess wastewater. *RSC Adv.* **2015**, *5*, 30801–30818. [CrossRef]

2. Zhang, Y.; Wu, B.; Xu, H.; Liu, H.; Wang, M.; He, Y.; Pan, B. Nanomaterials-Enabled water and wastewater treatment. *Nano Impact* **2016**, *3*, 22–39. [CrossRef]

3. Gitis, V.; Hankins, N. Water treatment chemicals: Trends and challenges. *J. Water Process Eng.* **2018**, *25*, 34–38. [CrossRef]

4. Hodges, B.C.; Cates, E.L.; Kim, J. Challenges and prospects of advanced oxidation water treatment processes using catalytic nanomaterial. *Nat. Nanotechnol.* **2018**, *13*, 642–650. [CrossRef] [PubMed]

5. Umar, K.; Haque, M.M.; Mir, N.A.; Muneer, M. Titanium dioxide-Mediated photocatalyzed mineralization of Two Selected organic pollutants in aqueous suspensions. *J. Adv. Oxid. Technol.* **2013**, *16*, 252–260.

6. Umar, K.; Ibrahim, M.N.M.; Ahmad, A.; Rafatullah, M. Synthesis of Mn-Doped TiO_2 by novel route and photocatalytic mineralization/intermediate studies of organic pollutants. *Res. Chem. Intermediat.* **2019**, *45*, 2927–2945. [CrossRef]

7. Baruah, S.; Khan, M.N.; Dutta, J. Perspectives and applications of nanotechnology in water treatment. *Environ. Chem. Lett.* **2016**, *14*, 1–14. [CrossRef]

8. Wu, Y.; Pang, H.; Liu, Y.; Wang, X.; Yu, S.; Fu, D.; Chen, J.; Wang, X. Environmental remediation of heavy metal ions by novel-Nanomaterials: A review. *Environ. Pollut.* **2019**, *246*, 608–620. [CrossRef]

9. Daer, S.; Kharraz, J.; Giwa, A.; Hasan, S.W. Recent applications of nanomaterials in water desalination: A critical review and future opportunities. *Desalination* **2015**, *367*, 37–48. [CrossRef]

10. Yaqoob, A.A.; Ibrahim, M.N.M. A Review Article of Nanoparticles; Synthetic Approaches and Wastewater Treatment Methods. *Int. Res. J. Eng. Technol.* **2019**, *6*, 1–7.

11. Tang, W.W.; Zeng, G.M.; Gong, J.L. Impact of humic/fulvic acid on the removal of heavy metals from aqueous solutions using nanomaterials: A review. *Sci. Total Environ.* **2014**, *468*, 1014–1027. [CrossRef]

12. Mir, N.A.; Haque, M.M.; Khan, A.; Umar, K.; Muneer, M.; Vijayalakshmi, S. Semiconductor mediated photocatalysed reaction of two selected organic compounds in aqueous suspensions of Titanium dioxide. *J. Adv. Oxid. Technol.* **2012**, *15*, 380–391. [CrossRef]

13. Umar, K. Water Contamination by Organic-Pollutants: TiO$_2$ Photocatalysis. In *Modern Age Environmental Problem and Remediation*; Oves, M., Khan, M.Z., Ismail, I.M.I., Eds.; Springer Nature: Basel, Switzerland, 2018; pp. 95–109.

14. Kalhapure, R.S.; Sonawane, S.J.; Sikwal, D.R. Solid lipid nanoparticles of clotrimazole silver complex: An efficient nano antibacterial against Staphylococcus aureus and MRSA. *Colloid Surf. B* **2015**, *136*, 651–658. [CrossRef] [PubMed]

15. Fang, X.; Li, J.; Li, X.; Pan, S.; Zhang, X.; Sun, X.; Han, J.S.W.; Wang, L. Internal pore decoration with polydopamine nanoparticle on polymeric ultrafiltration membrane for enhanced heavy metal removal. *Chem. Eng.* **2017**, *314*, 38–49. [CrossRef]

16. Sekoai, P.T.; Ouma, C.N.M.; Du Preez, S.P.; Modisha, P.; Engelbrecht, N.; Bessarabov, D.G.; Ghimire, A. Application of nanoparticles in biofuels: An overview. *Fuel* **2019**, *237*, 380–397. [CrossRef]

17. Briggs, A.M.; Cross, M.J.; Hoy, D.G.; Blyth, F.H.; Woolf, A.D.; March, L. Musculoskeletal Health Conditions Represent a Global Threat to Healthy Aging: A Report for the 2015 World Health Organization World Report on Ageing and Health. *Gerontologist* **2016**, *56*, 243–255. [CrossRef] [PubMed]

18. Umar, K.; Parveen, T.; Khan, M.A.; Ibrahim, M.N.M.; Ahmad, A.; Rafatullah, M. Degradation of organic pollutants using metal-Doped TiO$_2$ photocatalysts under visible light: A comparative study. *Desal. Water Treat.* **2019**, *161*, 275–282. [CrossRef]

19. Tang, K.; Gong, C.; Wang, D. Reduction potential, shadow prices, and pollution costs of agricultural pollutants in China. *Sci. Total Environ.* **2016**, *541*, 42–50. [CrossRef]

20. Richter, K.E.; Ayers, J.M. An approach to predicting sediment microbial fuel cell performance in shallow and deep water. *Appl. Sci.* **2018**, *8*, 2628. [CrossRef]

21. Sizmur, T.; Fresno, T.; Akgül, G.; Frost, H.; Jiménez, E.M. Review Biochar modification to enhance sorption of inorganics from water. *Bioresour. Technol.* **2017**, *246*, 34–47. [CrossRef]

22. Wang, J.; Wang, Z.; Carolina, L.Z.V.; Wolfson, J.M.; Pingtian, G. Review on the treatment of organic pollutants in water by ultrasonic technology. *Ultrasonics Sonochem.* **2019**, *55*, 273–278. [CrossRef] [PubMed]

23. Liu, C.; Hong, T.; Li, H.; Wang, L. From club convergence of per capita industrial pollutant emissions to industrial transfer effects: An empirical study across 285 cities in China. *Energy Policy* **2018**, *121*, 300–313. [CrossRef]

24. Bayoumi, T.A.; Saleh, H.M. Characterization of biological waste stabilized by cement during immersion in aqueous media to develop disposal strategies for phytomediated radioactive waste. *Prog. Nucl. Energy* **2018**, *107*, 83–89. [CrossRef]

25. Ma, H.; Guo, Y.; Qin, Y.; Li, Y.Y. Review Nutrient recovery technologies integrated with energy recovery by waste biomass anaerobic digestion. *Bioresour. Technol.* **2018**, *269*, 520–531. [CrossRef]

26. Longwane, G.H.; Sekoai, P.T.; Meyyappan, M.; Moothi, K. Review Simultaneous removal of pollutants from water using nanoparticles: A shift from single pollutant control to multiple pollutant control. *Sci. Total Environ.* **2019**, *656*, 808–833. [CrossRef]

27. Rajasulochana, P.; Preethy, V. Comparison on efficiency of various techniques in treatment of waste and sewage water-A comprehensive review. *Resour.-Effic. Technol.* **2016**, *4*, 175–184. [CrossRef]

28. Saravanan, R.; Gracia, F.; Stephen, A. Basic principles, mechanism, and challenges of photocatalysis. In *Nanocomposites for Visible Light-Induced Photocatalysis*; Springer, Cham: Berlin/Heidelberg, Germany, 2017; pp. 19–40.

29. Gomes, J.; Lincho, J.; Domingues, E.; Quinta-Ferreira, R.M.; Martins, R.C. N–TiO$_2$ photocatalysts: A review of their characteristics and capacity for emerging contaminants removal. *Water* **2019**, *11*, 373. [CrossRef]

30. Chen, W.; Liu, Q.; Tian, S.; Zhao, X. Exposed facet dependent stability of ZnO micro/nano crystals as a photocatalyst. *App. Surf. Sci* **2019**, *470*, 807–816. [CrossRef]

31. Ong, C.B.; Ng, L.Y.; Mohammad, A.W. A review of ZnO nanoparticles as solar photocatalysts: Synthesis, mechanisms and applications. *Renew. Sustain. Energy Rev.* **2018**, *81*, 536–551. [CrossRef]

32. Gómez-Pastora, J.; Dominguez, S.; Bringas, E.; Rivero, M.J.; Ortiz, I.; Dionysiou, D.D. Review and perspectives on the use of magnetic nanophotocatalysts (MNPCs) in water treatment. *Chem. Eng. J.* **2017**, *310*, 407–427. [CrossRef]

33. Umar, K.; Aris, A.; Parveen, T.; Jaafar, J.; Majid, Z.A.; Reddy, A.V.B.; Talib, J. Synthesis, Characterization of Mo and Mn doped Zno and their photocatalytic activity for the decolorization of two different chromophoric dyes. *Appl. Catal A* **2015**, *505*, 507–514. [CrossRef]

34. Loeb, S.K.; Alvarez, P.J.; Brame, J.A.; Cates, E.L.; Choi, W.; Crittenden, J.; Dionysiou, D.D.; Li, Q.; Li-Puma, G.; Quan, X.; et al. The technology horizon for photocatalytic water treatment: Sunrise or sunset? *Environ. Sci. Technol.* **2019**, *53*, 2937–2947. [CrossRef]

35. Reddy, A.V.B.; Jaafar, J.; Majid, Z.A.; Aris, A.; Umar, K.; Talib, J.; Madhavi, G. Relative efficiency comparison of carboxymethyl cellulose (cmc) stabilized fe0 and fe0/ag nanoparticles for rapid degradation of chlorpyrifos in aqueous solutions. *Dig. J. Nanomater. Bios.* **2015**, *10*, 331–340.

36. Samanta, H.S.; Das, R.; Bhattachajee, C. Influence of Nanoparticles for Wastewater Treatment-A Short Review. *Austin Chem. Eng.* **2016**, *3*, 1036–1045.

37. Qu, X.; Alvarez, P.J.; Li, Q. Applications of nanotechnology in water and wastewater treatment. *Water res* **2013**, *47*, 3931–3946. [CrossRef] [PubMed]

38. Sadegh, H.; Ali, G.A.M.; Gupta, V.K.; Makhlouf, A.S.H.; Nadagouda, M.N.; Sillanpaa, M.; Megiel, E. The role of nanomaterials as effective adsorbents and their applications in wastewater treatment. *J. Nanostructure Chem.* **2017**, *7*, 1–14. [CrossRef]

39. Raliya, S.R.; Avery, C.; Chakrabarti, S.; Biswas, P. Photocatalytic degradation of methyl orange dye by pristine TiO$_2$, ZnO, and graphene oxide nanostructures and their composites under visible light irradiation. *Appl. Nano Sci.* **2017**, *7*, 253–259. [CrossRef]

40. Liang, X.; Cui, S.; Li, H.; Abdelhady, A.; Wang, H.; Zhou, H. Removal effect on stormwater runoff pollution of porous concrete treated with nanometre titanium dioxide. *Transp. Res. D* **2019**, *73*, 34–45. [CrossRef]

41. Bhatia, D.; Sharma, N.R.; Singh, J.; Kanwar, R.S. Biological methods for textile dye removal from wastewater: A review. *Critcal Rev. Environ. Sci. Technol.* **2017**, *47*, 1836–1876. [CrossRef]

42. Sherman, J. Nanoparticulate Titanium Dioxide Coatings, and Processes for the Production and Use Thereof. U.S. Patent No, 6653356B2, 25 November 2003.

43. Ali, I.; Ghamdi, K.A.; Wadaani, F.T.A. Advances in iridium nano catalyst preparation, characterization and applications. *J. Mol. Liq.* **2019**, *280*, 274–284. [CrossRef]

44. Bhanvase, B.A.; Shende, T.P.; Sonawane, S.H. A review on grapheme-TiO$_2$ and doped grapheme-TiO$_2$ nanocomposite photocatalyst for water and wastewater treatment. *Environ. Technol. Rev.* **2017**, *6*, 1–14. [CrossRef]

45. Yamakata, A.; Junie Jhon, M.V. Curious behaviors of photogenerated electrons and holes at the defects on anatase, rutile, and brookite TiO$_2$ powders: A review. *J. Photochem. Photobiol C Phtotochem. Rev.* **2019**, *40*, 234–243. [CrossRef]

46. Chen, S.; Wang, Y.; Li, J.; Hu, Z.; Zhao, H.; Xie, W.; Wei, Z. Synthesis of black TiO$_2$ with efficient visible-light photocatalytic activity by ultraviolet light irradiation and low temperature annealing. *Mater Res. Bull.* **2018**, *98*, 280–287. [CrossRef]

47. Di Mauro, A.; Cantarella, M.; Nicotra, G.; Pellegrino, G.; Gulino, A.; Brundo, M.V.; Privitera, V.; Impellizzeri, G. Novel synthesis of ZnO/PMMA nanocomposites for photocatalytic applications. *Sci. Rep.* **2017**, *7*, 40–95. [CrossRef]

48. Hassan, A.F.; Elhadidy, H. Effect of Zr^{+4} doping on characteristics and sono catalytic activity of TiO$_2$/carbon nanotubes composite catalyst for degradation of chlorpyrifos. *J. Phys. Chem. Solids* **2019**, *129*, 180–187. [CrossRef]

49. Das, P.; Ghosh, S.; Ghosh, R.; Dam, S.; Baskey, M. *Madhuca longifolia* plant mediated green synthesis of cupric oxide nanoparticles: A promising environmentally sustainable material for wastewater treatment and efficient antibacterial agent. *J. Photochem. Photobiol.* **2018**, *189*, 66–73. [CrossRef]

50. Guya, N.; Cakar, S.; Ozacar, M. Comparison of palladium/zinc oxide photocatalysts prepared by different palladium doping methods for congo red degradation. *J. Colloid Interface Sci.* **2016**, *466*, 128–137. [CrossRef]

51. Bishoge, O.K.; Zhang, L.; Suntu, S.L.; Jin, H.; Zewde, A.A.; Qi, Z. Remediation of water and wastewater by using engineered nanomaterials: A review. *J. Environ. Sci. Heal A* **2018**, *53*, 537–554. [CrossRef]

52. Li, X.; Xia, T.; Xu, C.; Murowchick, J.; Chen, X. Synthesis and photoactivity of nanostructured CdS–TiO$_2$ composite catalysts. *Catal Today* **2014**, *225*, 64–73. [CrossRef]

53. Boyano, A.; Lázaro, M.J.; Cristiani, C.; Maldonado-Hodar, F.J.; Forzatti, P.; Moliner, R. A comparative study of V$_2$O$_5$/AC and V$_2$O$_5$/Al$_2$O$_3$ catalysts for the selective catalytic reduction of NO by NH$_3$. *Chem. Eng. J.* **2009**, *149*, 173–182. [CrossRef]

54. Serrà, A.; Zhang, Y.; Sepúlveda, B.; Gómez, E.; Nogués, J.; Michler, J.; Philippe, L. Highly reduced ecotoxicity of ZnO-Based micro/nanostructures on aquatic biota: Influence of architecture, chemical composition, fixation, and photocatalytic efficiency. *Water Res.* **2020**, *69*, 115210. [CrossRef]

55. Ameta, R.; Benjamin, S.; Ameta, A.; Ameta, S.C. Photocatalytic degradation of organic pollutants: A review. *Mater. Sci. Forum* **2013**, *734*, 247–272. [CrossRef]

56. Phokha, S.; Klinkaewnarong, J.; Hunpratub, S.; Boonserm, K.; Swatsitang, E.; Maensiri, S. Ferromagnetism in Fe-Doped MgO nanoparticles. *J. Mater. Sci. Mater. Electron.* **2016**, *27*, 33–39. [CrossRef]

57. Berekaa, M.M. Nanotechnology in wastewater treatment; influence of nanomaterials on microbial systems. *Int. J. Curr. Microbiol. App. Sci* **2016**, *5*, 713–726. [CrossRef]

58. Malik, A.; Hameed, S.; Siddiqui, M.J.; Haque, M.M.; Umar, K.; Khan, A.; Muneer, M. Electrical and optical properties of nickel-and molybdenum-doped titanium dioxide nanoparticle: Improved performance in dye-sensitized solar cells. *J. Mater. Eng. Perform.* **2014**, *23*, 3184–3192. [CrossRef]

59. Serrà, A.; Grau, S.; Gimbert-Suriñach, C.; Sort, J.; Nogués, J.; Vallés, E. Magnetically-Actuated mesoporous nanowires for enhanced heterogeneous catalysis. *App. Catal B Environ.* **2017**, *217*, 81–91. [CrossRef]

60. Ahmed, S.N.; Haider, W. Heterogeneous photocatalysis and its potential applications in water and wastewater treatment: A review. *Nanotechnology* **2018**, *29*, 342001. [CrossRef]

61. Kohtani, S.; Kawashima, A.; Miyabe, H. Stereoselective Organic Reactions in Heterogeneous Semiconductor Photocatalysis. *Front Chem.* **2019**, *7*, 630. [CrossRef]

62. Lekshmi, M.V.; Nagendra, S.S.; Maiya, M.P. Heterogeneous Photocatalysis for Indoor Air Purification: Recent Advances in Technology from Material to Reactor Modeling. In *Indoor Environmental Quality*; Springer: Berlin/Heidelberg, Germany, 2020; pp. 147–166.

63. Kanmani, S.; Sundar, K.P. Progression of Photocatalytic reactors and it's comparison: A Review. *Chem. Eng. Res. Des.* **2020**, *154*, 135–150.

64. Parrino, F.; Loddo, V.; Augugliaro, V.; Camera-Roda, G.; Palmisano, G.; Palmisano, L.; Yurdakal, S. Heterogeneous photocatalysis: Guidelines on experimental setup, catalyst characterization, interpretation, and assessment of reactivity. *Catal Rev.* **2019**, *61*, 163–213. [CrossRef]

65. Chong, M.N.; Jin, B.; Chow, C.W.; Saint, C. Recent developments in photocatalytic water treatment technology: A review. *Water Res.* **2010**, *44*, 2997–3027. [CrossRef]

66. Radhika, N.P.; Selvin, R.; Kakkar, R.; Umar, A. Recent advances in nano-photocatalysts for organic synthesis. *Arab. J. Chem.* **2019**, *12*, 4550–4578. [CrossRef]

67. Tahir, M.B.; Kiran, H.; Iqbal, T. The detoxification of heavy metals from aqueous environment using nano-photocatalysis approach: A review. *Environ. Sci. Pollut. Res.* **2019**, *26*, 10515–10528. [CrossRef] [PubMed]

68. Ciambelli, P.; La Guardia, G.; Vitale, L. Nanotechnology for green materials and processes. *Stud. Surf. Sci. Catal.* **2019**, *179*, 97–116.

69. Weng, B.; Qi, M.Y.; Han, C.; Tang, Z.R.; Xu, Y.J. Photocorrosion Inhibition of Semiconductor-Based Photocatalysts: Basic Principle, Current Development, and Future Perspective. *ACS Catal.* **2019**, *9*, 4642–4687. [CrossRef]

70. Rajabi, H.R.; Shahrezaei, F.; Farsi, M. Zinc sulfide quantum dots as powerful and efficient nanophotocatalysts for the removal of industrial pollutant. *J. Mater. Sci. Mater. Electron.* **2016**, *27*, 9297–9305. [CrossRef]

71. Mahmoodi, N.M.; Arami, M. Degradation and toxicity reduction of textile wastewater using immobilized titania nanophotocatalysis. *J. Photoch. Photobio. B* **2009**, *94*, 20–24. [CrossRef]

72. Van Gerven, T.; Mul, G.; Moulijn, J.; Stankiewicz, A. A review of intensification of photocatalytic processes. *Chem. Eng. Process. Process Intensif.* **2007**, *46*, 781–789. [CrossRef]

73. Lin, W.Y.; Wang, Y.; Wang, S.; Tseng, H.R. Integrated microfluidic reactors. *Nano Today* **2009**, *4*, 470–481. [CrossRef]

74. Wang, N.; Zhang, X.; Wang, Y.; Yu, W.; Chan, H.L. Microfluidic reactors for photocatalytic water purification. *Lab on a Chip* **2014**, *14*, 1074–1082. [CrossRef]

75. Umar, K.; Dar, A.A.; Haque, M.M.; Mir, N.A.; Muneer, M. Photocatalysed decolourization of two textile dye derivatives, Martius Yellow and Acid Blue 129 in UV-irradiated aqueous suspensions of Titania. *Desal. Water Treat.* **2012**, *46*, 205–214. [CrossRef]

76. Rasolevandi, T.; Naseri, S.; Azarpira, H.; Mahvi, A.H. Photo-Degradation of dexamethasone phosphate using UV/Iodide process: Kinetics, intermediates, and transformation pathways. *J. Mol. Liq.* **2019**, *295*, 111703–111710. [CrossRef]

77. Rayaroth, M.P.; Aravind, U.K.; Aravindakumar, C.T. Photocatalytic degradation of lignocaine in aqueous suspension of TiO2 nanoparticles: Mechanism of degradation and mineralization. *J. Environ. Chem. Eng.* **2018**, *6*, 3556–3564. [CrossRef]

78. Jurado-Sánchez, B.; Wang, J. Micromotors for environmental applications: A review. *Environ. Sci. Nano* **2018**, *5*, 1530–1544. [CrossRef]

79. Moo, J.G.S.; Pumera, M. Chemical energy powered nano/micro/macromotors and the environment. *Chem.–A Eur. J.* **2015**, *21*, 58–72. [CrossRef]

80. Pacheco, M.; López, M.Á.; Jurado-Sánchez, B.; Escarpa, A. Self-Propelled micromachines for analytical sensing: A critical review. *Anal. Bioanal. Chem.* **2019**, *411*, 6561–6573. [CrossRef]

81. Chi, Q.; Wang, Z.; Tian, F.; You, J.A.; Xu, S. A review of fast bubble-Driven micromotors powered by biocompatible fuel: Low-Concentration fuel, bioactive fluid and enzyme. *Micromachine* **2018**, *9*, 537. [CrossRef]

82. García-Torres, J.; Serrà, A.; Tierno, P.; Alcobé, X.; Vallés, E. Magnetic propulsion of recyclable catalytic nanocleaners for pollutant degradation. *ACS Appl. Mater. Interfac.* **2017**, *9*, 23859–23868. [CrossRef]

83. Pourrahimi, A.M.; Pumera, M. Multifunctional and self-Propelled spherical Janus nano/micromotors: Recent advances. *Nanoscale* **2018**, *10*, 16398–16415. [CrossRef]

84. Eskandarloo, H.; Kierulf, A.; Abbaspourrad, A. Nano-and micromotors for cleaning polluted waters: Focused review on pollutant removal mechanisms. *Nanoscale* **2017**, *9*, 13850–13863. [CrossRef]

85. Gao, W.; Uygun, A.; Wang, J. Hydrogen-Bubble-Propelled Zinc-Based Microrockets in Strongly Acidic Media. *J. Am. Chem. Soc.* **2012**, *134*, 897–900. [CrossRef] [PubMed]

86. Zhao, G.; Stuart, E.J.E.; Pumera, M. Enhanced Diffusion of Pollutants by Self-Propulsion. *Phys. Chem. Chem. Phys.* **2011**, *13*, 12755–12757. [CrossRef] [PubMed]

87. Seah, T.H.; Zhao, G.; Pumera, M. Surfactant Capsules Propel Interfacial Oil Droplets: An Environmental Cleanup Strategy. *Chem. Plus. Chem.* **2013**, *78*, 395–397.

88. Kagan, D.; Calvo-Marzal, P.; Balasubramanian, S.; Sattayasamitsathit, S.; Manesh, K.M.; Flechsig, G.U.; Wang, J. Chemical Sensing Based on Catalytic Nanomotors: Motion-Based Detection of Trace Silver. *J. Am. Chem. Soc.* **2009**, *131*, 12082–12083. [CrossRef]

89. Orozco, J.; Cortes, A.; Cheng, G.; Sattayasamitsathit, S.; Gao, W.; Feng, X.; Shen, Y.; Wang, J. Molecularly Imprinted Polymer-Based Catalytic Micromotors for Selective Protein Transport. *J. Am. Chem. Soc.* **2013**, *135*, 5336–5339. [CrossRef]

90. Wu, J.; Balasubramanian, S.; Kagan, D.; Manesh, K.M.; Campuzano, S.; Wang, J. Motion-Based DNA Detection Using Catalytic Nanomotors. *Nat. Commun.* **2010**, *1*, 36. [CrossRef]

91. Orozco, J.; Cheng, G.; Vilela, D.; Sattayasamitsathit, S.; Vazquez-Duhalt, R.; Valdes-Ramirez, G.; Pak, O.S.; Escarpa, A.; Kan, C.; Wang, J. Micromotor-Based High-Yielding Fast Oxidative Detoxification of Chemical Threats. *Angew. Chem. Int. Ed.* **2013**, *52*, 13276–13279. [CrossRef]

92. Soler, L.; Magdanz, V.; Fomin, V.M.; Sanchez, S.; Schmidt, O.G. Self-Propelled Micromotors for Cleaning Polluted Water. *ACS Nano* **2013**, *7*, 9611–9620. [CrossRef]

93. Guix, M.; Orozco, J.; Garcia, M.; Gao, W.; Sattayasamitsathit, S.; Merkoci, A.; Escarpa, A.; Wang, J. Superhydrophobic Alkanethiol-Coated Microsubmarines for Effective Removal of Oil. *ACS Nano* **2012**, *6*, 4445–4451. [CrossRef]

94. Gao, W.; Pei, A.; Dong, R.; Wang, J. Catalytic Iridium-Based Janus Micromotors Powered by Ultralow Levels of Chemical Fuels. *J. Am. Chem. Soc.* **2014**, *136*, 2276–2279. [CrossRef]

95. Dey, K.K.; Bhandari, S.; Bandyopadhyay, D.; Basu, S.; Chattopadhyay, A. The pH Taxis of an Intelligent Catalytic Microbot. *Small* **2013**, *9*, 1916–1920. [CrossRef] [PubMed]

96. Soler, L.; Sánchez, S. Catalytic nanomotors for environmental monitoring and water remediation. *Nanoscale* **2014**, *6*, 7175–7182. [CrossRef] [PubMed]

97. Li, J.; Singh, V.V.; Sattayasamitsathit, S.; Orozco, J.; Kaufmann, K.; Dong, R.; Gao, W.; Jurado-Sanchez, B.; Fedorak, Y.; Wang, J. Water-Driven micromotors for rapid photocatalytic degradation of biological and chemical warfare agents. *ACS Nano* **2014**, *8*, 11118–11125. [CrossRef] [PubMed]

98. Orozco, J.; Mercante, L.A.; Pol, R.; Merkoçi, A. Graphene-Based Janus micromotors for the dynamic removal of pollutants. *J. Mater. Chem. A* **2016**, *4*, 3371–3378. [CrossRef]

99. Li, J.; Chang, H.; Ma, L.; Hao, J.; Yang, R.T. Low-Temperature selective catalytic reduction of NOx with NH3 over metal oxide and zeolite catalysts-A review. *Catal Today* **2011**, *175*, 147–156. [CrossRef]

100. Pourrahimi, A.M.; Villa, K.; Ying, Y.; Sofer, Z.; Pumera, M. ZnO/ZnO2/Pt Janus Micromotors Propulsion Mode Changes with Size and Interface Structure: Enhanced Nitroaromatic Explosives Degradation under Visible Light. *ACS Appl. Mater. Interfaces* **2018**, *10*, 42688–42697. [CrossRef]

101. Srivastava, S.K.; Guix, M.; Schmidt, O.G. Wastewater mediated activation of micromotors for efficient water cleaning. *Nano Lett.* **2016**, *16*, 817–821. [CrossRef]

102. Liu, J.; Hong, C.; Shi, X.; Nawar, S.; Werner, J.; Huang, G.; Ye, M.; Weitz, D.A.; Solovev, A.A.; Mei, Y. Hydrogel Microcapsules with Photocatalytic Nanoparticles for Removal of Organic Pollutants. *Environ. Sci. Nano* **2020**. [CrossRef]

103. Gao, W.; D'Agostino, M.; Garcia-Gradilla, V.; Orozco, J.; Wang, J. Multi-fuel driven janus micromotors. *Small* **2013**, *9*, 467–471. [CrossRef]

104. Pourrahimi, A.M.; Liu, D.; Ström, V.; Hedenqvist, M.S.; Olsson, R.T.; Gedde, U.W. Heat treatment of ZnO nanoparticles: New methods to achieve high-Purity nanoparticles for high-voltage applications. *J. Mater. Chem. A* **2015**, *3*, 17190–17200. [CrossRef]

105. Ge, H.; Chen, X.; Liu, W.; Lu, X.; Gu, Z. Metal-Based Transient Micromotors: From Principle to Environmental and Biomedical Applications. *Chem.–Asian J.* **2019**, *14*, 2348–2356. [CrossRef]

106. Zhang, Q.; Dong, R.; Wu, Y.; Gao, W.; He, Z.; Ren, B. Light-Driven Au-WO3@ C Janus micromotors for rapid photodegradation of dye pollutants. *ACS App. Mater. Interface* **2017**, *9*, 4674–4683. [CrossRef] [PubMed]

107. Zhang, Y.; Yuan, K.; Zhang, L. Micro/nanomachines: From functionalization to sensing and removal. *Adv. Mater. Technol.* **2019**, *4*, 1800636–1800658. [CrossRef]

108. Wang, H.; Khezri, B.; Pumera, M. Catalytic DNA-Functionalized self-Propelled micromachines for environmental remediation. *Chem.* **2016**, *1*, 473–481. [CrossRef]

109. Zhang, B.; Huang, G.; Wang, L.; Wang, T.; Liu, L.; Di, Z.; Liu, X.; Mei, Y. Rolled-Up Monolayer Graphene Tubular Micromotors: Enhanced Performance and Antibacterial Property. *Chem.–Asian J.* **2019**, *14*, 2479–2484. [CrossRef] [PubMed]

110. Wang, L.; Song, H.; Yuan, L.; Li, Z.; Zhang, P.; Gibson, J.K.; Zheng, L.; Wang, H.; Chai, Z.; Shi, W. Effective Removal of Anionic Re (VII) by Surface-Modified Ti2CT x MXene Nanocomposites: Implications for Tc (VII) Sequestration. *Environ. Sci. Technol.* **2019**, *53*, 3739–3747. [CrossRef] [PubMed]

111. Gao, W.; Dong, R.; Thamphiwatana, S.; Li, J.; Gao, W.; Zhang, L.; Wang, J. Artificial micromotors in the mouse's stomach: A step toward in vivo use of synthetic motors. *ACS Nano* **2015**, *9*, 117–123. [CrossRef]

112. Safdar, M.; Simmchen, J.; Jänis, J. Correction: Light-Driven micro-and nanomotors for environmental remediation. *Environ. Sci. Nano* **2017**, *4*, 2235. [CrossRef]

113. Fu, P.P.; Xia, Q.; Hwang, H.M.; Ray, P.C.; Yu, H. Mechanisms of nanotoxicity: Generation of reactive oxygen species. *J. Food Drug Anal.* **2014**, *22*, 64–75. [CrossRef]

114. Ying, Y.; Pumera, M. Micro/Nanomotors for Water Purification. *Chem–Eur. J.* **2019**, *25*, 106–121. [CrossRef]

115. Fernández-Medina, M.; Ramos-Docampo, M.A.; Hovorka, O.; Salgueiriño, V.; Städler, B. Recent Advances in Nano-and Micromotors. *Adv. Funct. Mater.* **2020**, 1908283–19082299. [CrossRef]

116. Jhaveri, J.H.; Murthy, Z.V.P. A comprehensive review on anti-Fouling nanocomposite membranes for pressure driven membrane separation processes. *Desalination* **2016**, *379*, 137–154. [CrossRef]

117. Hogen-Esch, T.; Pirbazari, M.; Ravindran, V.; Yurdacan, H.M.; Kim, W. High Performance Membranes for Water Reclamation Using Polymeric and Nanomaterials. U.S. Patent No. 20160038885A, 29 October 2019.

118. Waduge, P.; Larkin, J.; Upmanyu, M.; Kar, S.; Wanunu, M. Programmed Synthesis of Freestanding Graphene Nanomembrane Arrays. *Nano Microphone* **2015**, *11*, 597–603.

119. Bassyouni, M.; Abdel-Aziz, M.H.; Zoromba, M.S.; Abdel Hamid, S.M.S.; Drioli, E. A review of polymeric nanocomposite membranes for water purification. *J. Ind. Eng. Chem.* **2019**, *73*, 19–46. [CrossRef]

120. Zahid, M.; Rashid, A.; Akram, S.; Rehan, Z.A.; Razzaq, W. A Comprehensive Review on Polymeric Nano-Composite Membranes for Water Treatment. *J. Membr. Sci. Technol.* **2018**, *8*, 179–190. [CrossRef]

121. Saleh, A.; Parthasarathy, P.; Irfan, M. Advanced functional polymer nanocomposites and their use in water ultra-purification. *Trends Environ Anal.* **2019**, *24*, 67–78. [CrossRef]

122. Kochkodan, V.; Hilal, N. A comprehensive review on surface modified polymer membranes for biofouling mitigation. *Desalination* **2015**, *356*, 187–207. [CrossRef]

123. Ronen, A.; Duan, W.; Wheeldon, I.; Walker, S.; Jassby, D. Microbial Attachment Inhibition through Low-Voltage Electrochemical Reactions on Electrically Conducting Membranes. *Environ. Sci. Technol.* **2015**, *49*, 12741–12750. [CrossRef]

124. Yin, J.; Yang, Y.; Hu, Z.; Deng, B. Attachment of silver nanoparticles (AgNPs) onto thin-Film composite (TFC) membranes through covalent bonding to reduce membrane biofouling. *J. Membr. Sci.* **2013**, *441*, 73–82. [CrossRef]

125. Zhang, M.; Field, R.W.; Zhang, K. Biogenic silver nanocomposite polyethersulfone UF membranes with antifouling properties. *J. Membr. Sci.* **2014**, *471*, 274–284. [CrossRef]

126. Hirata, K.; Watanabe, H.; Kubo, W. Nanomembranes as a substrate for ultra-thin lightweight devices. *Thin Solid Film.* **2019**, *676*, 8–11. [CrossRef]

127. Gopalakrishnan, I.; Samuel, S.R.; Sridharan, K. nanomaterials-Based adsorbents for water and waste water treatment. *Emerg. Nanotechnol. Environ. Sustain.* **2018**, *6*, 89–98.

128. Liu, S.; Wang, Y.; Zhou, Z.; Hana, W.; Li, J.; Shen, J.; Wang, I. Improved degradation of the aqueous flutriafol using a nanostructure microporous PbO_2 as reactive electrochemical membrane. *Electrochim. Acta* **2017**, *253*, 357–367. [CrossRef]

129. Shetti, N.P.; Bukkitgar, S.D.; Reddy, K.R.; Aminabhavi, T.M. Nanostructured titanium oxide hybrids-Based electrochemical biosensors for healthcare applications. *Colloids Surf. B Biointerfaces* **2019**, *178*, 385–394. [CrossRef] [PubMed]

130. Kunduru, R.K.; Kovsky, M.N.; Rajendra, S.F.; Pawar, P.; Basu, A.; Domb, A.J. Nanotechnology for water purification: Applications of nanotechnology methods in wastewater treatment. *Water Purif.* **2017**, *10*, 33–74.

131. Ibrahim, R.K.; Hayyan, M.; Al-saadi, M.A.; Hayyan, A.; Ibrahim, S. Environmental application of nanotechnology; air, soil, and water. *Environ. Sci. Pollut. R* **2016**, *23*, 13754–13788. [CrossRef] [PubMed]

132. Mekaru, H.; Lu, J.; Tamanoi, F. Development of mesoporous silica-Based nanoparticles with controlled release capability for cancer therapy. *Adv. Drug Deliv. Rev.* **2015**, *95*, 40–49. [CrossRef]

133. Jawed, A.; Saxena, V.; Pandey, L.M. Engineered nanomaterials and their surface functionalization for the removal of heavy metals: A review. *J. Wat. Process. Eng.* **2020**, *33*, 101009. [CrossRef]

134. Umar, K.; Aris, A.; Ahmad, H.; Parveen, T.; Jaafar, J.; Majid, Z.A.; Reddy, A.V.B.; Talib, J. Synthesis of visible light active doped TiO_2 for the degradation of organic pollutants-Methylene blue and glyphosate. *J. Anal. Sci. Technol.* **2016**, *7*, 29–36. [CrossRef]

135. Kumar, M.; Patil, P.; Kim, G.D. Marine microorganisms for synthesis of metallic nanoparticles and their biomedical applications. *Coll. Surface B* **2018**, *172*, 487–495.

136. Al-Ghouti, M.A.; Kaabi, M.A.A.; Ashfaq, M.Y.; Dana, D.A. Produced water characteristics, treatment and reuse: A review. *J. Water Proc. Eng.* **2019**, *28*, 222–239. [CrossRef]

137. Ahn, Y.Y.; Yun, E.T.; Seo, J.W.; Lee, C.; Kim, S.H.; Lee, J. Activation of peroxymonosulfate by surface loaded Nobel metal nanoparticles for oxidative degradation of organic compounds. *Environ. Sci. Technol.* **2016**, *50*, 10187–10197. [CrossRef] [PubMed]

138. Abdullah, N.; Yusof, N.; Lau, W.J.; Jaafar, J.; Ismail, A.F. Review Recent trends of heavy metal removal from water/wastewater by membrane technologies. *J. Ind. Eng. Chem.* **2019**, *76*, 17–38. [CrossRef]

139. Bhat, A.H.; Rehman, W.U.; Khan, I.U.; Ahmad, S.; Ayoub, M.; Usmani, M.A. Nanocomposite membrane for environmental remediation. In *Polymer-Based Nanocomposites for Energy and Environmental Applications*; Jawaid, M., Khan, M.M., Eds.; Woodhead Publishing: Cambridge, UK, 2018; pp. 407–440.

140. Muntha, S.T.; Kausar, A.; Siddiq, M. Advances in polymeric nanofiltration membrane: A review. *Polym.-Plast Technol. Eng.* **2017**, *56*, 841–856. [CrossRef]

141. Rashidi, H.R.; Sulaiman, N.M.N.; Hashim, N.A.; Hassan, C.R.C.; Ramli, M.R. Synthetic reactive dye wastewater treatment by using nano-Membrane filtration. *Desalin. Water Treat.* **2015**, *55*, 86–95. [CrossRef]

142. Belloň, T.; Polezhaev, P.; Vobecká, L.; Slouka, Z. Fouling of a heterogeneous anion-Exchange membrane and single anion-Exchange resin particle by ssdna manifests differently. *J. Membr. Sci.* **2019**, *572*, 619–631. [CrossRef]

143. Naeem, F.; Naeem, S.; Zhao, Z.; Shu, G.Q.; Zhang, J.; Mei, Y.; Huang, G.S. Atomic layer deposition synthesized ZnO nanomembranes: A facile route towards stable supercapacitor electrode for high capacitance. *J. Power Source* **2020**, *451*, 227740. [CrossRef]

144. Feng, C.; Khulbe, K.C.; Matsuura, T. Recent progress in the preparation, characterization, and applications of nanofibers and nanofiber membranes via electrospinning/interfacial polymerization. *J. Appl. Polym. Sci.* **2010**, *115*, 756–776. [CrossRef]

145. Esfahani, M.R.; Aktij, S.A.; Dabaghian, Z.; Firouzjaei, M.D.; Rahimpour, A.; Eke, J.; Escobar, I.C.; Abolhassani, M.; Greenlee, L.F.; Esfahani, A.R.; et al. Nanocomposite membranes for water separation and purification: Fabrication, modification, and applications. *Sep. Purif. Technol.* **2019**, *213*, 465–499. [CrossRef]

146. Tang, C.; Wang, Z.; Petrinić, I.; Fane, A.G.; Hélix-Nielsen, C. Biomimetic aquaporin membranes coming of age. *Desalination* **2015**, *368*, 89–105. [CrossRef]

147. Cornwell, D.J.; Smith, D.K. Expanding the scope of gels–Combining polymers with low-MOLECULAR-Weight gelators to yield modified self-Assembling smart materials with high-Tech applications. *Mater. Horiz.* **2015**, *2*, 279–293. [CrossRef]

148. Asatekin, A.; Menniti, A.; Kang, S.; Elimelech, M.; Morgenroth, E.; Mayes, A.M. Antifouling nanofiltration membranes for membrane bioreactors from self-assembling graft copolymers. *J. Membr. Sci.* **2006**, *285*, 81–89. [CrossRef]

149. Ying, Y.; Ying, W.; Li, Q.; Meng, D.; Ren, G.; Yan, R.; Peng, X. Recent advances of nanomaterial-Based membrane for water purification. *App. Mater. Today* **2017**, *7*, 144–158. [CrossRef]

150. Montemagno, C.; Schmidt, J.; Tozzi, S. Biomimetic Membranes. U.S. Patent No. 20040049230, 11 March 2004.

151. Salim, W.; Ho, W.S.W. Recent developments on nanostructured polymer-Based membranes. *Curr. Opin. Chem. Eng.* **2015**, *8*, 76–82. [CrossRef]

152. Yu, L.; Ruan, S.; Xu, X.; Zou, R.; Hu, J. Review One-Dimensional nanomaterial-Assembled macroscopic membranes for water treatment. *Nano Today* **2017**, *17*, 79–95. [CrossRef]

153. Fuwad, A.; Ryu, H.; Malmstadt, N.; Kim, S.M.; Jeon, T.J. Biomimetic membranes as potential tools for water purification: Preceding and future avenues. *Desalination* **2019**, *458*, 97–115. [CrossRef]

154. Shen, Y.X.; Saboe, P.O.; Sines, I.T.; Erbakan, M.; Kumar, M. Biomimetic membranes: A review. *J. Membr. Sci.* **2014**, *454*, 359–381. [CrossRef]

155. Peng, F.; Xu, T.; Wu, F.; Ma, C.X.; Liu, Y.; Li, J.; Zhao, B.; Mao, C. Novel biomimetic enzyme for sensitive detection of superoxide anions. *Talanta* **2017**, *174*, 82–91. [CrossRef]

156. Giwa, A.; Hasan, S.W.; Yousaf, A.; Chakraborty, S.; Johnson, D.J.; Hilal, N. Biomimetic membranes: A critical review of recent progress. *Desalination* **2017**, *420*, 403–424. [CrossRef]

157. Diallo, M.S. Water Treatment by Dendrimer-Enhanced Filtration. U.S. Patent No. 2009/0223896, 10 September 2009.

158. Sahebi, S.; Sheikhi, M.; Ramavandi, B. A new biomimetic aquaporin thin film composite membrane for forward osmosis: Characterization and performance assessment. *Desalin. Water Treat.* **2019**, *148*, 42–50. [CrossRef]

159. Perez, T.; Pasquini, D.; Lima, A.F.; Rosa, E.V.; Sousa, M.H.; Cerqueira, D.A.; Morais, L.C. Efficient removal of lead ions from water by magnetic nanosorbents based on manganese ferrite nano particles capped with thin layers of modified biopolymers. *J. Environ. Chem. Eng.* **2019**, *7*, 802–892. [CrossRef]

160. Manikam, M.K.; Halim, A.A.; Hanafiah, M.M.; Krishnamoorthy, R.R. Removal of ammonia nitrogen, nitrate, phosphorus and COD from sewage wastewater using palm oil boiler ash composite adsorbent. *Desal. Water Treat.* **2019**, *149*, 23–30. [CrossRef]

161. Charee, S.W.; Aravinthan, V.; Erdei, L.; Raj, W.S. Use of macadamia nut shell residues as magnetic nanosorbents. *Int. Biodeter. Biodegr.* **2017**, *124*, 276–287.

162. Rodovalho, F.L.; Capistrano, G.; Gomes, J.A.; Sodre, F.F.; Chaker, J.A.; Campos, A.F.C.; Bakuzis, A.F.; Sousa, A.H. Elaboration of magneto-Thermally recyclable nanosorbents for remote removal of toluene in contaminated water using magnetic hyperthermia. *Chem. Eng. J.* **2016**, *15*, 725–732. [CrossRef]

163. Sun, X.; Liu, Z.; Zhang, G.; Qiu, G.; Zhong, N.; Wu, L.; Cai, D.; Wu, Z. Reducing the pollution risk of pesticide using nano networks induced by irradiation and hydrothermal treatment. *J. Environ. Sci. Health C* **2015**, *50*, 901–907. [CrossRef] [PubMed]

164. Lee, X.J.; Foo, L.P.Y.; Tan, K.W.; Hassell, D.G.; Lee, L.Y. Evaluation of carbon-Based nanosorbents synthesized by ethylene decomposition on stainless steel substrates as potential sequestrating materials for nickel ions in aqueous solution. *J. Environ. Sci.* **2012**, *24*, 1559–1568. [CrossRef]

165. Kyzas, G.Z.; Matis, K.A. Review Nanoadsorbents for pollutants removal: A review. *J. Mol. Liq.* **2015**, *203*, 159–168. [CrossRef]

166. Wang, Y.; Zhang, Y.; Hou, C.; Liu, M. Mussel-Inspired synthesis of magnetic polydopamine–Chitosan nanoparticles as bio sorbent for dyes and metals removal. *J. Taiwan Inst. Chem. E* **2016**, *61*, 292–298. [CrossRef]

167. Krstić, V.; Pesovski, T.U.B. A review on adsorbents for treatment of water and wastewaters containing copper ions. *Chem. Eng. Sci.* **2018**, *192*, 273–287. [CrossRef]

168. Unuabonah, E.I.; Taubert, A. Review article Clay-Polymer nanocomposites (CPNs): Adsorbents of the future for water treatment. *Appl. Clay Sci.* **2014**, *99*, 83–92. [CrossRef]

169. Yadav, V.B.; Gadi, R.; Kalra, S. Clay based nanocomposites for removal of heavy metals from water: A review. *J. Environ. Manag.* **2019**, *232*, 803–817. [CrossRef] [PubMed]

170. Vunain, E.; Mishra, A.K.; Mamba, B.B. Dendrimers, mesoporous silicas and chitosan-Based nanosorbents for the removal of heavy-Metal ions: A review. *Int. J. Biolog. Macromol.* **2016**, *86*, 570–586. [CrossRef] [PubMed]

171. Khajeh, M.; Laurent, S.; Dastafkan, K. Nanoadsorbents: Classification, preparation, and applications (with emphasis on aqueous media). *Chem. Rev.* **2013**, *113*, 7728–7768. [CrossRef] [PubMed]

172. Zhang, Y.H.; Hu, C.Z.; Liu, F.; Yuan, Y.; Wu, H.; Li, A. Effects of ionic strength on removal of toxic pollutants from aqueous media with multifarious adsorbents: A review. *Sci. Total Environ.* **2019**, *646*, 265–279. [CrossRef]

173. Kobielska, P.; Howarth, A.J.; Farha, O.K.; Nayak, S. Review Metal–Organic frameworks for heavy metal removal from water. *Coord. Chem. Rev.* **2018**, *358*, 92–107. [CrossRef]

174. Quesada, H.B.; Baptista, A.T.A.; Cusioli, L.F.; Seibert, D.; Bezerra, C.O.; Bergamasco, R. Surface water pollution by pharmaceuticals and an alternative of removal by low-Cost adsorbents: A review. *Chemosphere* **2019**, *222*, 766–780. [CrossRef]

175. Brandao, D.; Liebana, S.; Pividori, M.I. Multiplexed detection of foodborne pathogens based on magnetic particles. *New Biotechnol.* **2015**, *8*, 76–82. [CrossRef]

176. Jones, D.; Caballero, S.; Pardo, G.D. Bioavailability of nanotechnology-Based bioactive and nutraceuticals. *Adv. Food Nutr. Res.* **2019**, *1*, 2–9.

177. Wang, S.; Lu, W.; Tovmachenko, O.; Rai, U.S.; Yu, H.; Ray, P.C. Challenge in Understanding Size and Shape Dependent Toxicity of Gold Nanomaterials in Human Skin Keratinocytes. *Chem. Phys. Lett.* **2008**, *463*, 145–149. [CrossRef]

178. Takahashi, H.; Niidome, Y.; Niidome, T.; Kaneko, K.; Kawasaki, H.; Yamada, S. Modification of gold nanorods using phosphatidylcholine to reduce cytotoxicity. *Langmuir* **2006**, *22*, 2–5. [CrossRef]

179. Karlsson, H.L.; Cronholm, P.; Gustafsson, J.; Möller, L. Copper oxide nanoparticles are highly toxic: A comparison between metal oxide nanoparticles and carbon nanotubes. *Chem. Res. Toxicol.* **2008**, *21*, 1726–1732. [CrossRef] [PubMed]

180. Ajdary, M.; Moosavi, M.A.; Rahmati, M.; Falahati, M.; Mahboubi, M.; Mandegary, A.; Jangjoo, I.S.; Mohammad inejad, R.; Varma, R.S. Health Concerns of Various Nanoparticles: A Review of Their in Vitro and in Vivo Toxicity. *Nanomater* **2018**, *1*, 634. [CrossRef] [PubMed]

181. Ray, P.C.; Hongtao, Y.; Peter, P. Toxicity and Environmental Risks of Nanomaterials: Challenges and Future Needs. *J. Environ. Sci. Health C Environ. Carcinog. Ecotoxicol. Rev.* **2009**, *27*, 1–35. [CrossRef] [PubMed]

182. Taju, G.; Majeed, S.A.; Nambi, K.; Hameed, A.S. In vitro assay for the toxicity of silver nanoparticles using heart and gill cell lines of catla catla and gill cell line of labeo rohita. *Comp. Biochem. Physiol. C Toxicol. Pharmacol.* **2014**, *161*, 41–52. [CrossRef]

183. Jia, G.; Wang, H.; Yan, L.; Wang, X.; Pei, R.; Yan, T.; Zhao, Y.; Guo, X. Cytotoxity of carbon nanomaterials: Single-Wall nanotube, multi-wall nanotube, and fullerene. *Environ. Sci. Technol.* **2005**, *39*, 1378–1383. [CrossRef]

184. Baig, N.; Ihsanullah, M.; Sajjid, T.A.; Saleh, A. Graphene-Based adsorbents for the removal of toxic organic pollutants: A review. *J. Environ. Manag.* **2019**, *244*, 370–382. [CrossRef]

185. Mansoori, G.A.; Bastami, T.R.; Ahmadpour, A.; Eshaghi, Z. Environmental application of nanotechnology. *Annu. Rev. Nano Res.* **2008**, *2*, 439–493.

186. Ersan, G.; Apul, O.G.; Perreault, F.; Karanfil, T. Adsorption of organic contaminants by graphene nanosheets: A review. *Water Res.* **2017**, *126*, 385–398. [CrossRef]

187. Kumar, V.; Kumar, P.; Pounara, A.; Vellingiri, K.; Kim, K.H. Nanomaterials for the sensing of narcotics: Challenges and opportunities. *Trends Anal Chem.* **2018**, *2*, 84–115. [CrossRef]

188. Heijden, V.D. Developments and challenges in the manufacturing, characterization and scale-Up of energetic nanomaterials-A review. *Chem. Eng. J.* **2018**, *350*, 939–948. [CrossRef]

189. Adeleye, A.S.; Conway, J.R.; Garner, K.; Huang, Y.; Su, Y.A.; Keller, A. Engineered nanomaterials for water treatment and remediation: Costs, benefits, and applicability. *Chem. Eng. J.* **2016**, *286*, 640–662. [CrossRef]

190. Pandey, N.; Shukla, S.K.; Singh, N.B. Water purification by polymer nanocomposites: An overview. *Nanocomposites* **2017**, *3*, 47–66. [CrossRef]

191. Santhosh, C.; Velmurugan, V.P.; Jacob, G.; Jeong, S.K.; Grace, N.A.; Bhatnagar, A. Role of nanomaterials in water treatment applications: A review. *Chem. Eng. J.* **2016**, *306*, 1116–1137. [CrossRef]

192. Tate, J.E.; Burton, A.H.; Pinto, C.B. Global, Regional, and National Estimates of Rotavirus Mortality in Children <5 Years of Age, 2000–2013. *Clin. Infect. Dis.* **2016**, *62*, 96–105. [CrossRef] [PubMed]

193. Prathna, T.C.; Sharma, S.K.; Kennedy, M. Review Nanoparticles in household level water treatment: An overview. *Sep. Purif. Technol.* **2018**, *199*, 260–270.

194. Umar, K.; Haque, M.M.; Muneer, M.; Harada, T.; Matsumura, M. Mo, Mn and La doped TiO$_2$: Synthesis, characterization and photocatalytic activity for the decolourization of three different chromophoric dyes. *J. Alloy Compd.* **2013**, *578*, 431–438. [CrossRef]

195. Soppe, A.I.A.; Heijman, S.G.J.; Gensburger, I.; Shantz, A.; Halem, D.V.; Kroesbergen, J.; Wubbels, G.H.; Smeets, P.W.M.H. Critical parameters in the production of ceramic pot filters for household water treatment in developing countries. *J. Water Health* **2014**, *13*, 587–599. [CrossRef]

196. Bushra, R.; Shahadat, M.; Ahmad, A.; Nabi, S.A.; Umar, K.; Muneer, M.; Raeissia, A.S.; Owais, M. Synthesis, characterization, antimicrobial activity and applications of composite adsorbent for the analysis of organic and inorganic pollutants. *J. Hazard. Mater.* **2014**, *264*, 481–489. [CrossRef]

197. Nor, N.A.M.; Jaafar, J.; Othman, M.H.D.; Rahman, M.A. A review study of nanofibers in photocatalytic process for wastewater treatment. *J. Teknologi.* **2013**, *65*, 83–88. [CrossRef]

198. Zhang, Y.Z.; Wang, X.; Feng, Y.; Li, J.; Lim, C.T.; Ramakrishna, S. Coaxial electrospinning of (fluorescein isothiocyanate-Conjugated bovine serum albumin)-Encapsulated poly (ε-caprolactone) nanofibers for sustained release. *Biomacromolecule* **2006**, *7*, 1049–1057. [CrossRef]

199. Chuang, C.S.; Wang, M.K.; Ko, C.H.; Ou, C.C.; Wu, C.H. Removal of benzene and toluene by carbonized bamboo materials modified with TiO$_2$. *Bioresour. Technol.* **2008**, *99*, 954–958. [CrossRef] [PubMed]

200. Wu, C.H.; Chang, H.W.; Chern, J.M. Basic dye decomposition kinetics in a photocatalytic slurry reactor. *J. Hazard Mater.* **2006**, *137*, 336–343. [CrossRef]

201. Paek, S.M.; Jung, H.; Lee, Y.J.; Park, M.; Hwang, S.J.; Choy, J.H. Exfoliation and reassembling route to mesoporous titania nanohybrids. *Chem. Mater.* **2006**, *18*, 1134–1140. [CrossRef]

202. Fu, W.; Yang, H.; Chang, L.; Li, M.; Zou, G. Anatase TiO$_2$ nanolayer coating on strontium ferrite nanoparticles for magnetic photocatalyst. *Colloid Surf. A Physicochem. Eng. Asp.* **2006**, *289*, 47–52. [CrossRef]

203. Hamadanian, M.; Reisi-Vanani, A.; Majedi, A. Synthesis, characterization and effect of calcination temperature on phase transformation and photocatalytic activity of Cu,S-codoped TiO$_2$ nanoparticles. *Appl. Surf. Sci.* **2010**, *256*, 1837–1844. [CrossRef]

204. Yu, C.; Cai, D.; Yang, K.; Yu, J.C.; Zhou, Y.; Fan, C. Sol–gel derived S,I-codoped mesoporous TiO$_2$ photocatalyst with high visible-Light photocatalytic activity. *J. Phys. Chem. Solids* **2010**, *71*, 1337–1343. [CrossRef]

205. Umar, M.; Aziz, H.A. Photocatalytic degradation of organic pollutants in water. *Org. Pollut.-Monit. Risk Treat.* **2013**, *8*, 196–197.

206. Dimapilis, E.A.S.; Hsu, C.S.; Mendoza, R.M.O.; Lu, M.C. Zinc oxide nanoparticles for water disinfection. *Sustain. Environ.* **2018**, *28*, 47–56. [CrossRef]

207. Sultana, S.; Rafiuddin; Khan, M.Z.; Umar, K.; Ahmed, A.S.; Shahadat, M. SnO$_2$-SrO based nanocomposites and their photocatalytic activity for the treatment of organic pollutants. *J. Mol. Struct.* **2015**, *1098*, 393–399. [CrossRef]

208. Faisal, M.; Tariq, M.A.; Khan, A.; Umar, K.; Muneer, M. Photochemical reactions of 2, 4-dichloroaniline and 4-nitroanisole in aqueous suspension of titanium dioxide. *Sci. Adv. Mater.* **2011**, *3*, 269–275. [CrossRef]

209. He, D.; Sun, Y.; Xin, L.; Feng, J. Aqueous tetracycline degradation by non-Thermal plasma combined with nano-TiO$_2$. *Chem. Eng. J.* **2014**, *258*, 18–25. [CrossRef]

210. Mir, N.A.; Khan, A.; Umar, K.; Muneer, M. Photocatalytic Study of a Xanthene Dye Derivative, Phloxine B in Aqueous Suspension of TiO2: Adsorption Isotherm and Decolourization Kinetics. *Energy Environ. Focus* **2013**, *2*, 208–216. [CrossRef]

211. Dar, A.A.; Umar, K.; Mir, N.A.; Haque, M.M.; Muneer, M.; Boxall, C. Photocatalysed degradation of an herbicide derivative, Dinoterb, in aqueous suspension. *Res. Chem. Intermediat.* **2011**, *37*, 567–578. [CrossRef]
212. Schlosser, D. Biotechnologies for Water Treatment. In *Advanced Nano-Bio Technologies for Water and Soil Treatment*; Springer, Cham: Berlin/Heidelberg, Germany, 2020; pp. 335–343.
213. Lu, F.; Astruc, D. Nanocatalysts and other nanomaterials for water remediation from organic pollutants. *Coord. Chem. Rev.* **2020**, *408*, 213180. [CrossRef]

MDPI
St. Alban-Anlage 66
4052 Basel
Switzerland
Tel. +41 61 683 77 34
Fax +41 61 302 89 18
www.mdpi.com

Water Editorial Office
E-mail: water@mdpi.com
www.mdpi.com/journal/water